Karl Theiler

The House Mouse

Development and Normal Stages from Fertilization to 4 Weeks of Age

With 335 Figures

Springer-Verlag
New York Heidelberg Berlin 1972

Dr. KARL THEILER
Professor of Anatomy, University of Zürich, CH-8006 Zürich

ISBN 0-387-05940-7 Springer-Verlag New York · Heidelberg · Berlin
ISBN 3-540-05940-7 Springer-Verlag Berlin · Heidelberg · New York

Printed by Universitätsdruckerei H. Stürtz AG, Würzburg

Preface

In the course of our studies in developmental genetics, the need of an accurate time-table of the development of inbred mice became apparent. There are "tables of normal stages" for some species, but for the mouse, in spite of its widespread use in biological research, there is no complete table. The description of mouse development by R. Rugh [5] contains a wealth of information up to (but it stops at) 16 days of gestation, and it is especially addressed to students.

The "equivalent ages in mouse and human embryos" by Otis and Brent [4] is a valuable brief summary. The careful studies of Snell and Stevens in Green's [2] book, as well as those of Lewis and Wright [16] cover only limited periods in development.

This atlas is intended to provide a broader documentation of mouse development than is presently available. Hybrids between two inbred strains were bred under standardized conditions to produce a series of embryos, fetuses, and young animals of precisely known age. Nonetheless, within a single litter, there may be differences in fertilization time which can result in developmental variations of up to 20 hours. For this reason, copulations were timed within a period ± two hours.

I am indebted to Miss Mirjam Erni for the technical preparation of embryos, slides and tables, and to Miss Marcelle Ernst for preparing the rough-draft.

The English translation has been thoroughly revised by Dr. Leroy C. Stevens from the Jackson Laboratory, Bar Harbor, Maine, whose continued interest was an invaluable help to me. My thanks are also due to Dr. Margareth Green of the Jackson Laboratory, as well to all those, in Switzerland and abroad, who encouraged me in this work.

The preparation of this book has been financed in part by the generous support of the Emil Barell-Stiftung of Hoffmann-LaRoche AG.

Zürich, autumn 1972 Karl Theiler

Contents

Introduction

The external form typical of each age group is represented by profile photographs and there is a short description of characteristic features. Some organs such as the eye, ear, pituitary and pineal gland, are particularly useful for determining stages in development. These are described in more detail. To complete the picture of the structural organization as a whole, graphic reconstructions were made.
Any subdivision of development into stages must necessarily be arbitrary. I have tried to use as a guide the classification of Streeter [6] in his "Developmental horizons of human embryos", and I have pointed out equivalent human-mouse stages [4] whenever possible. For older human embryos, I have used the age data of Olivier and Pineau [Bull. Ass. Anat. *47*, 573–576 (1962)]. This classification is restricted to initial organogenesis, and precise comparisons in staging between human and mouse cannot be made. Some phases of development are faster in the mouse than in the human, and some are slower. For practical reasons, in this text, both stage number and age in days will be given together. The chapter numbers also correspond to the stage numbers.
The reference list is subdivided under various headings. It is brief and emphasizes recent investigations. No references on the rat are included, and the author apologizes for the inevitable incompleteness.

Materials and Methods

To obtain hybrid embryos, females of our inbred C57BL/6 stock were crossed, at the age of 3–5 months, with CBA males. Usually first litters were used, but a few second litters were included.
Strains: C57BL/6 Th, F? + 20, transferred to Zurich in 1960 from The Jackson Laboratory, Bar Harbor, Maine. CBA F? + 10, also received in 1960 from The Jackson Laboratory. The nomenclature of inbred strains is explained in: J. Heredity *54*, 159–162, 1963, and in Green [2].

Procedure

Animals were placed in a room in which the normal light-dark sequence was reversed for at least 4 days. Three or four females were placed with a male in total darkness. Copulation was determined by the presence of a vaginal plug. The middle of the "artificial night" was designated as day 0 of pregnancy. It is not considered identical with ovulation time. A late ovulation may result in considerable retardation in development. On the other hand, an early ovulation does not advance development, because the second maturation division is completed only after fertilization.
For each period several litters were examined. The most advanced embryos were regarded as representative of the respective age groups. If by chance all litters of one group stem from

especially early or late ovulations, a difference from the "developmental curve" would result. When this was observed, additional litters were obtained for this group. For the first 3 days, successive stages were designated a full day apart. Days 4 to 12 were designated a half day apart. The 12-hour staging regularly revealed an overlapping of developmental phases.

Some specimens were fixed in Bouin's solution, 4% formol or Carnoy's solution, and imbedded in paraffin. Some early embryos were fixed in OsO_4 and embedded in Methacrylate to prepare thin sections.

The serial sections were routinely stained with Hematoxylin-Eosin (H.-E.). Additional information was gained by the use of periodic-acid-Schiff-reaction. Older specimens were stained with azocarmine-anilinblue. Cleared skeletons were stained with alizarin-red S or methylgreen.

Stage 1 One-celled Egg

1–20 Hours

Human equivalent
Horizon I, one-celled egg

Stage 1 begins with *fertilization.* It invariably occurs in the ampulla tubae, the dilated uppermost loop of the coiled oviduct (Fig. 1). *The eggs*, after ovulation, are in the metaphase stage of the second maturation division (Fig. 2). They are surrounded by follicle cells which tend to clump together (Fig. 4). In each ampulla, there are 3 to 5 ova.

After 6 hours the eggs are still encircled by several layers of follicle cells. Sometimes these show mitotic cell divisions, side by side with pycnotic nuclei. Some ova are already fertilized. As an example, the specimen KT 980 contained 9 ova of which 2 could definitely be regarded as fertilized. No first polar body was seen in any of these cells.

Evidently the first polar body soon becomes cytolized. After ovulation, it is usually no longer visible. However, it may persist in exceptional cases, for instance, as in specimen KT 791, a 2-celled egg of 24 hours.

After 10 hours the eggs are still located in the ampulla tubae. There are fewer follicle cells surrounding them, and many are pycnotic. The percentage of fertilized ova has considerably increased: for example all of 8 egg cells in specimen KT 982 were fertilized, and all have emitted the second polar body (Fig. 5). All except one contained a male pronucleus (Fig. 6). CHANG [11] observed that not even 50% of the eggs in this stage were fertilized.

After 20 hours all follicle cells have disappeared. The eggs are now located between the first and second loop of the oviduct (KT 972). The male pronucleus approaches the female pronucleus and starts mitosis (Figs. 7 and 8). The *zona pellucida* shrinks considerably after fixation in Bouin's solution, so that the overall diameter, in the fixed state, amounts to only 55 to 60 microns.

Overripe eggs are from time to time encountered in the oviduct (KT 995, KT 966). Within the first 10 hours after copulation, they can be easily recognized by the lack of surrounding follicle cells. In one case, a "nude" single egg contained 4 pronuclei of different sizes. Sometimes overripe eggs seem to initiate division, and they consist of several loosely adhering cells with only faintly staining nuclei. Similar degenerating blastomeres often occur in the ovary, within atretic follicles. They may arise by "spontaneous parthenogenesis". Artificial parthenogenesis has been described recently [15].

Material	Age	Content
KT 993/94	2 h	3 eggs in meiosis
KT 966	5 h	degenerating eggs
KT 995	5 h	6 eggs, some degenerating
KT 979	6 h	degenerating eggs
KT 980	6 h	9 eggs with female pronucleus, some fertilized
KT 981	10 h	8 eggs in second meiotic division
KT 971/72	20 h	8 eggs: one 2-celled, 1 triploid, 1 degenerating
KT 977/78	20 h	7 eggs: 6 definitely fertilized, with pronuclei and polocytes

Figs. 1–8: Begin of development, first day

Fig. 1 Overall picture: ovary–oviduct–uterine horn.
A Ampulla tubae, *Cl* freshly ruptured follicle, *I* infundibulum, *O* ostium uterinum tubae, projecting in uterine lumen.
KT 972. 22.5:1

Fig. 2 Tubal egg in 2nd meiotic division, not fertilized, in ampulla.
KT 981. 720:1

Fig. 3 Spermatozoa, epididymal smear, iron-hematoxylin.
R Cytoplasmic droplet at end of mid piece. 1300:1

Fig. 4 Ampulla tubae with 2 fertilized eggs and surrounding follicle cells. 135:1

Fig. 5 Detail, showing sperm head under zona pellucida.
Opposite to it: polar body "*P*", in telophase.
KT 982, 10 h. 720:1

Fig. 6 Male and female pronucleus.
KT 982, 10 h. 720:1

Fig. 7 Male pronucleus approaches and starts mitosis.
Polar body "*P*" is detached.
KT 972, 20 h. 900:1

Fig. 8 Both nuclear membranes dissolved. Amphimixis.
KT 978, 20 h. 900:1

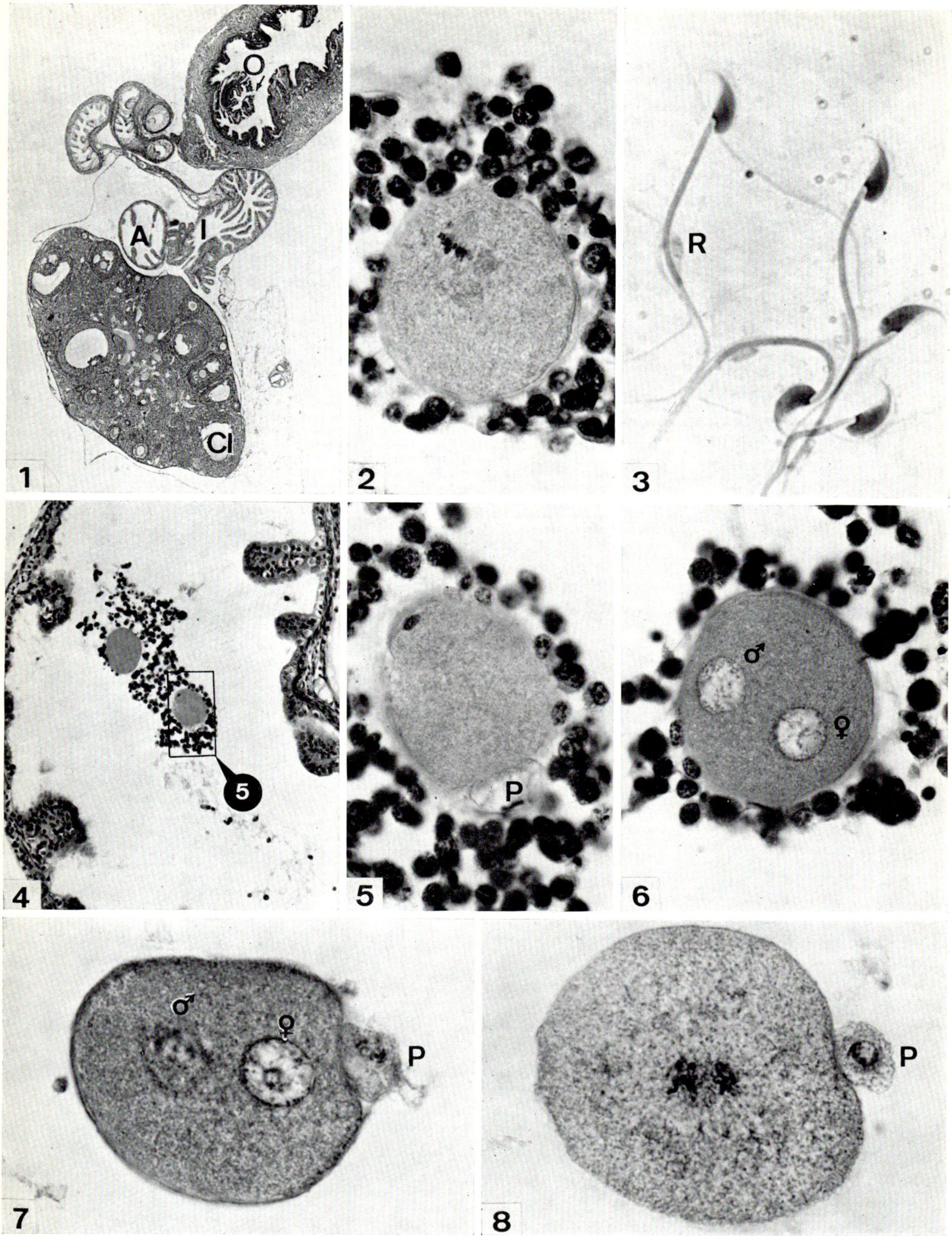
1
O
A
I
Cl
2
3
R
4
5
5
P
6
♂
♀
7
♂
♀
P
8
P

Horizon II/a
Segmenting egg

Stage 2 Beginning of Segmentation

20–24 Hours

Within 24 hours the first cleavage division is completed. The eggs are near the exit of the ampulla, between the first and the second loop of the oviduct (Fig. 9). This second loop is said to exhibit peristaltic contractions which aid ova transport [5]. *The first cleavage* yields two cells of about equal size, with finely granulated cytoplasm. Their large spherical nuclei contain 4–5 nucleoli which are surrounded by a small border of chromatin. The egg and zona pellucida shrink considerably after Bouin's fixation. The overall-diameter is 48 to 64 microns, including the zona. After fixation with OsO_4 there is much less shrinkage and the zona appears in sections as a distinct thick ring (Figs. 10 and 11), with an overall-diameter of 83 microns. In the fresh, unfixed state, eggs of this stage are 80–100 microns in diameter.
The second polar body is tangentially cut in Fig. 12. The first polar body, visible in another plane of section, is located about 90 degrees from the second, and is in metaphase. The nucleus of the second polocyte is typically small and has peripheral chromatin (Fig. 10).
The corpus luteum has slightly enlarged cells which form irregular trabeculae separated by invading capillaries. As an exception, in Fig. 9 a distinct central blood coagulum is visible.
Spermatozoa are visible in the oviduct and uterus up to 20 hours after copulation (KT 975). There are also few leucocytes and other free cells. Thereafter, spermatozoa disappear completely.

Material	Age	Content
KT 975/76	20 h	4 fertilized eggs: one 2-celled, in the process of cleavage, one degenerating
KT 790/91	24 h	4 fertilized eggs: all 2-celled, at lower end of first loop of oviduct
KT 729	45 h	eggs not fertilized, some exhibiting second meiotic division, some degenerating

Figs. 9–12: Begin of segmentation, 20 h

Fig. 9 Overall picture: ovary–oviduct.
Drawing (right) shows location of eggs in oviduct (*arrow*); *Cl* fresh corpus luteum.
Bouin, H.-E.
KT 791. 40:1

Fig. 10 Two-celled egg with polar body "*P*".
Phase contrast, Fixation OsO_4.
KT 790. 270:1

Fig. 11 Two-celled egg.
In nucleus "*N*" several small nucleoli are visible. Phase contrast. Fixation OsO_4.
KT 790. 270:1

Fig. 12 Two-celled egg, higher magnification. Bouin, H.-E.
The zona pellucida obviously thinned in comparison to Figs. 10 and 11.
KT 791. 580:1

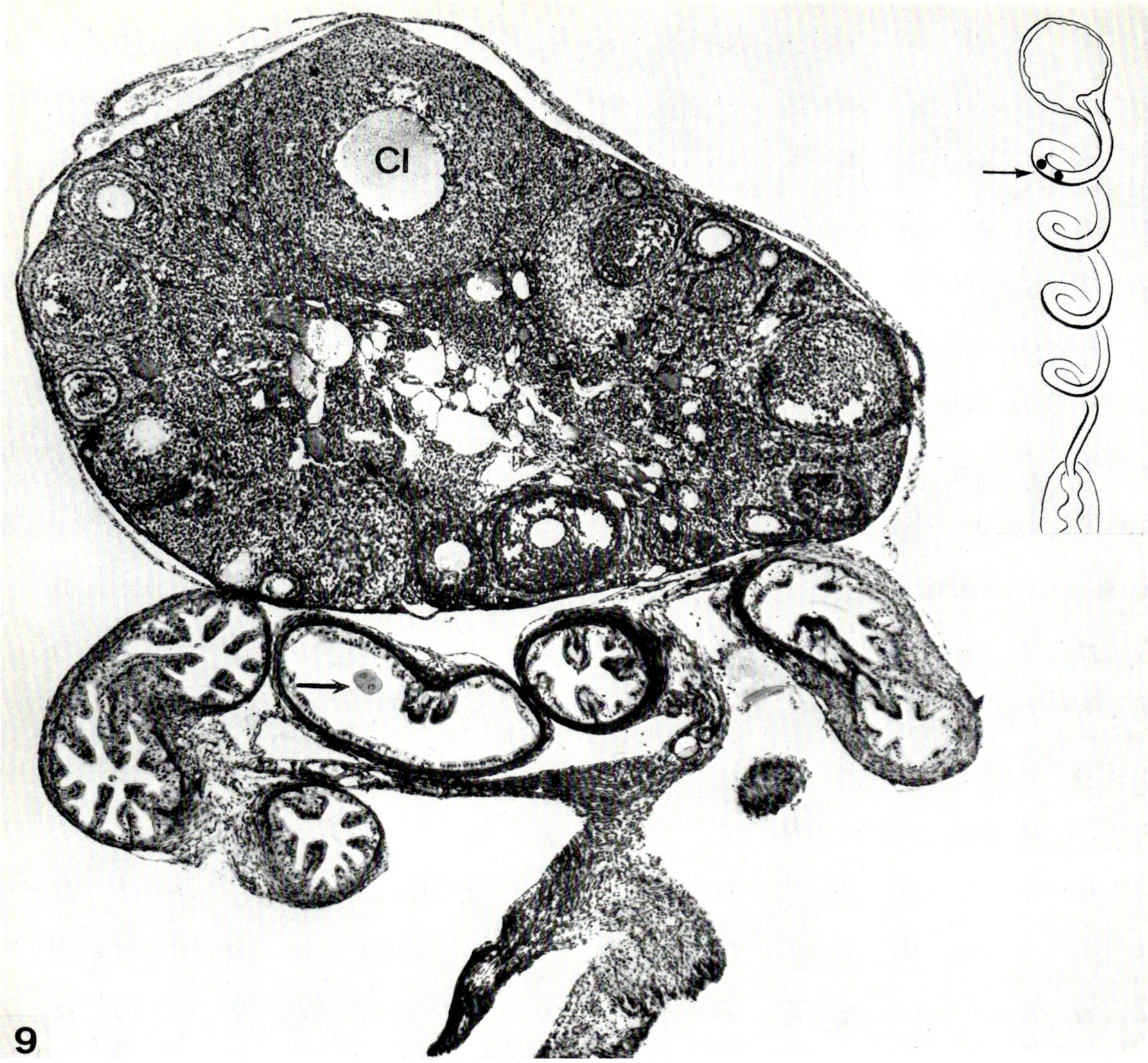

9

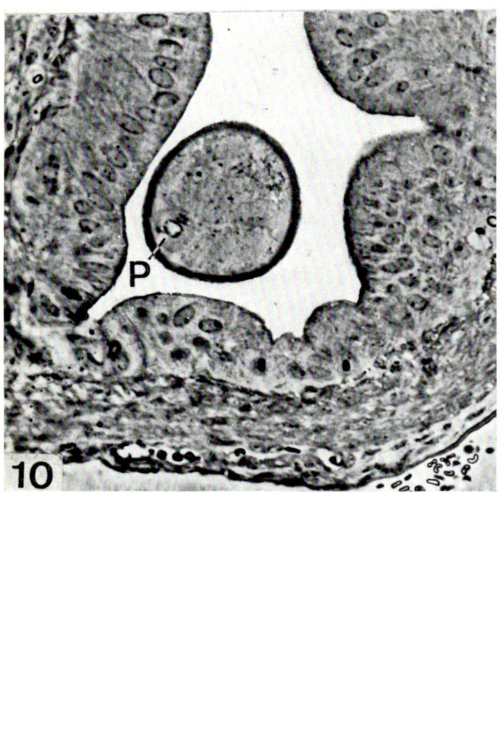

10

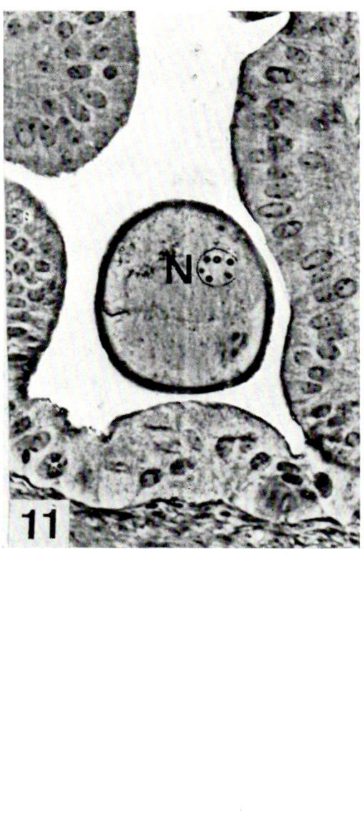

11

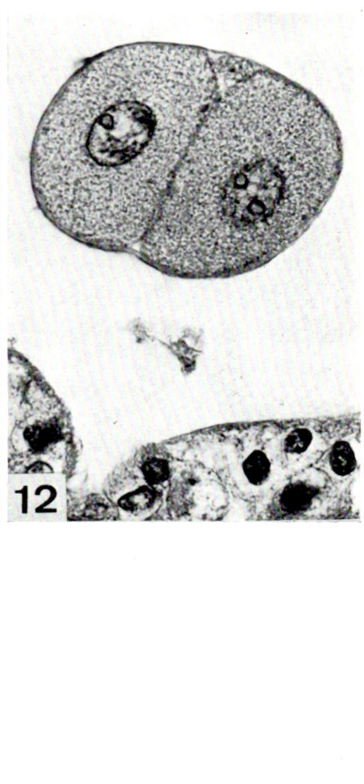

12

Horizon II/b
Segmenting egg

Stage 3 Segmenting Egg

2 Days, Morula

After 52–53 hours the embryos are composed of *2 to 16 cells*. This difference in degree of development must have existed in the previous stage, and is probably a result of different fertilization times.

The *blastomeres* are of unequal size, and some are dividing mitotically. The cytoplasm is finely granulated throughout and the nuclei are rather small. On the other hand, nucleoli enlarge considerably, and their number decreases. In the two-celled egg, 4 or 5 small nucleoli are visible in each cell. In the 4-celled stage, there are only 2 or 3.

A segmentation cavity is not yet visible.

The diameter of the morulae after Bouin fixation is 70 microns (KT 845), including the shrunken *zona pellucida*. The eggs are still in clusters. They have advanced to the lower half of the oviduct.

The polocytes are still present, even in the 16-celled morulae. Their nuclei appear pycnotic.

Corpora lutea measure 700–750 microns. The central cavity is filled by connective tissue and contains little blood and fibrin. Rarely leucocytes are encountered (Fig. 15).

Spermatozoa are no longer visible in the uterus or the oviduct.

Material	Age	Eggs
KT 846	52 h	1 one-celled and 3 four-celled eggs
KT 845	53 h	4 six-celled, 1 nine-celled, and 1 sixteen-celled egg

Figs. 13–17: Cleavage, 53 h

Fig. 13 Corpus luteum, overall picture, Bouin, H.-E.
Right: drawing showing location of eggs in oviduct.
KT 845. 125:1

Fig. 14 Detail of Fig. 13
showing the connective tissue organization of the central region. 360:1

Fig. 15 Detail of the adjacent section.
Arrow: leucocytes appear in connection with the central reorganization. 720:1

Fig. 16 Segmenting egg of 9 cells, with polar body "*P*" in oviduct, Bouin, H.-E.
KT 845. 270:1

Fig. 17 Segmenting egg of 9 cells, median section.
Zona pellucida extremely thin (fixation effect).
KT 845. 760:1

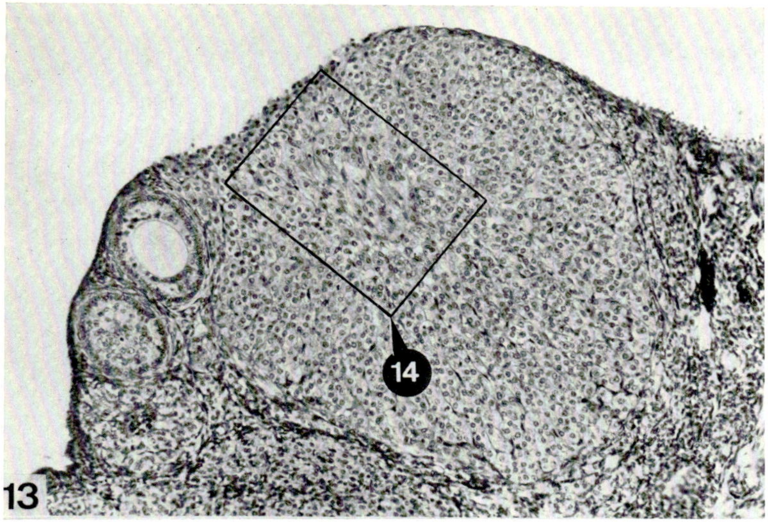
13
14

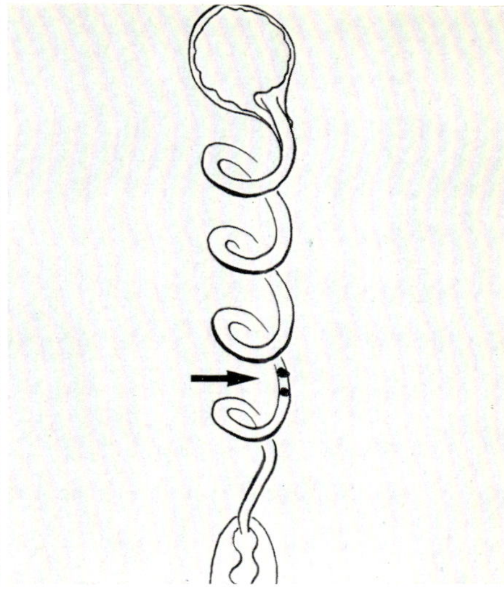

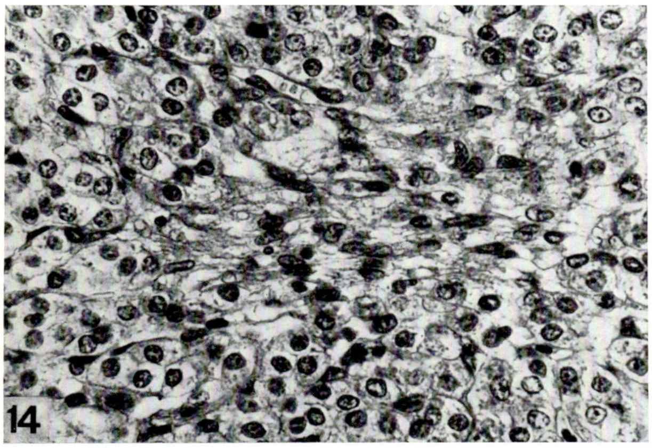
14

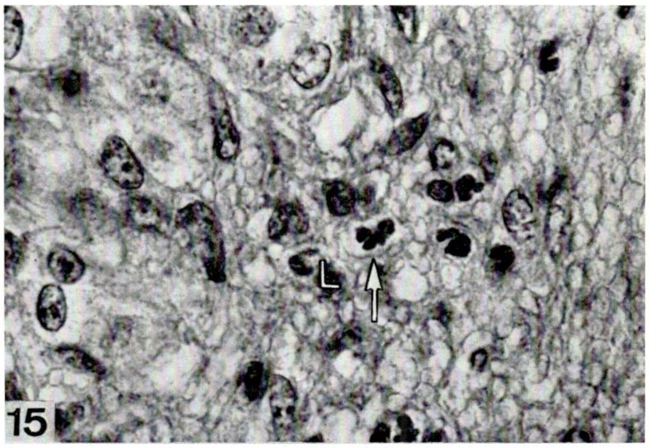
L
15

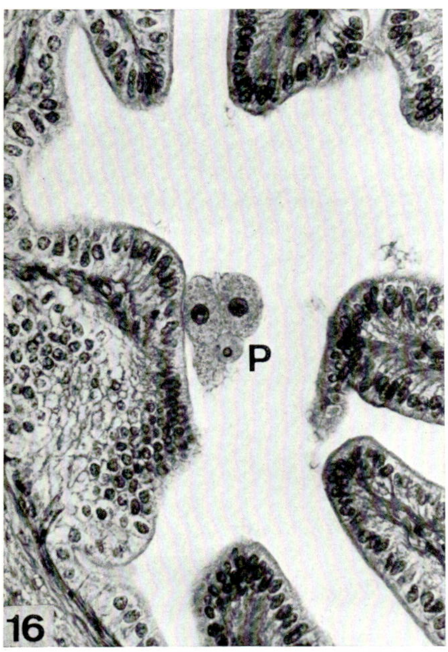
P
16

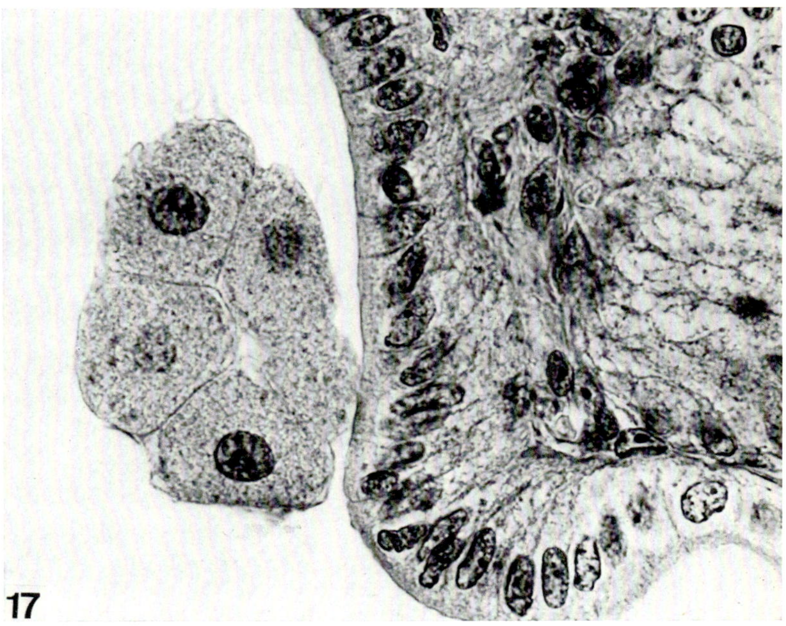
17

Stage 4 Advanced Segmentation

3 Days

Horizon II/c
Segmenting egg

On the third day, the eggs are in the uterus [17]. After 71 hours, one was found in the pars intramuralis tubae (KT 786). The eggs were composed of 16–25 cells after 69–71 hours.
The *blastomeres* are not quite equal in size. The cytoplasm is coursely granulated, with spherical or rod-like eosinophilic inclusions. The nucleoli are extraordinarily large and may be one-third of the nuclear diameter.
Out of 15 nuclei of specimen KT 934:

10 showed a single big nucleolus,
3 showed two nucleoli of unequal size,
2 showed several small nucleoli.

The trophoblast cannot yet be distinguished with certainty from the embryoblast. The segmentation cavity forms very rapidly (78 hours), and makes it possible to distinguish the embryoblast from the trophoblast. The 25-celled egg lacks a distinct cavity, but the 35-celled egg (78 hours) has a large eccentrically placed lumen.
The *diameter* of the morulae has apparently not changed. After fixation in Bouin's solution or Carnoy's solution (which is suitable for a PAS-reaction) it was 62–70 microns. The *zona pellucida* stains intensively red with the PAS reaction.
Polocytes are sometimes still distinct. Their nuclei are pycnotic and slightly larger than the nucleolus of a blastomere. The cytoplasm is PAS-negative.
The *distances* between the eggs have increased, and the eggs are irregularly spaced.
The *corpora lutea* are more intensely vascularized than in the previous stage.

Material	Age	Morulae
KT 786	71 h	18-celled, in pars intramuralis tubae
KT 777	69 h	22-celled, 5 of which are in mitosis, in uterus
KT 934	69 h	1 showing 16 cells, with distinct polocyte 1 showing 18 cells 1 showing 20 cells, 2 of which are in mitosis 1 showing 25 cells, 7 of which are in mitosis
KT 989/90	78 h	2 morulae with 33 and 34 cells, in uterus 1 blastocyst, 37 cells
KT 991/92	78 h	1 morula, 31 cells 4 blastocysts: 35, 36, 42 and 43 cells

Figs. 18–22: Cleavage, 69 h

Fig. 18 Low magnification. Uterine horn, cross sectioned, Bouin, H.-E. KT 934. 54:1

Fig. 19 Detail of Fig. 18 with 16-celled morula. Drawing (right) shows location of eggs (*arrow*). 360:1

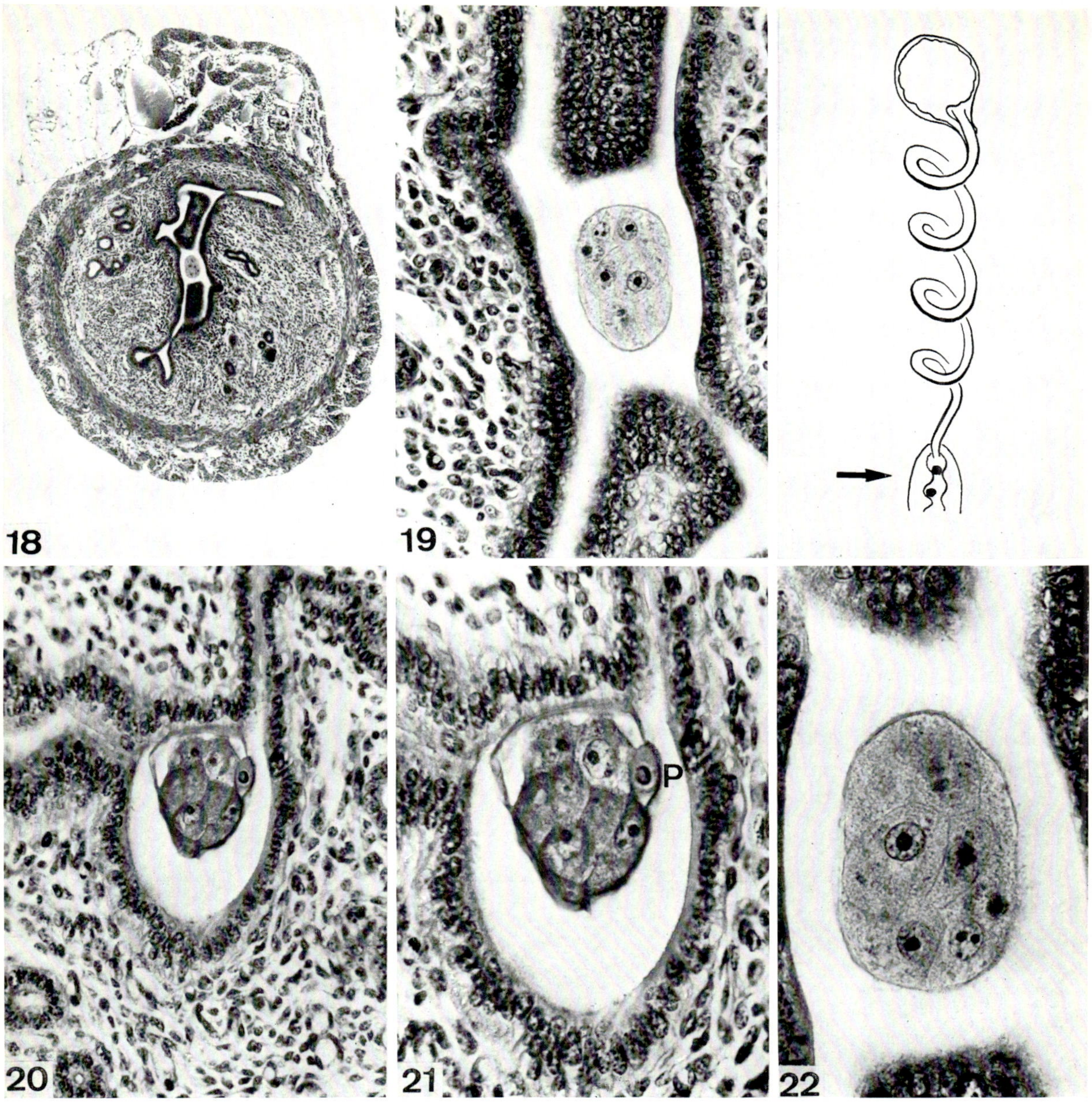

Fig. 20 Morula of 21 cells, Carnoy, PAS.
Distinct zona pellucida, partly separated from blastomeres because of shrinkage. Beginning formation of a crypt in uterine lumen. No deciduous reaction yet.
KT 934. 360:1

Fig. 21 Morula of 21 cells.
Polar body "*P*" with pycnotic nucleus (not to be confounded with a nucleolus).
KT 934a. 360:1

Fig. 22 Morula of 16 cells, Bouin, H.-E.
Nuclei containing one big or two small nucleoli. Zona pellucida very thin (fixation effect).
KT 934. 720:1

Horizon III
Free blastocyst

Stage 5 Blastocyst

4 Days

After one hundred hours, almost all blastocysts have arrived in the uterus. The only exception was specimen KT 969, where the right oviduct contained 4 probably decaying blastocysts. None had reached the right horn of the uterus. The other uterine horn had 3 normal blastocysts, and the adjoining tube was empty.

The eggs are distinctly spaced along the entire uterus, apparently free within the lumen, sometimes in a crypt. In most cases, the wall of the blastocyst was probably already in close contact with the uterine epithelium, and is secondarily separated by fixation shrinkage (Fig. 29).

The *blastocyst* is clearly separated into embryoblast and trophoblast. The trophoblastic cells are flattened, and form a single-layered epithelium. They are said to absorb amino acids, perhaps influenced by hormones [19].

The embryoblastic cells are cuboidal and clustered at the embryonic pole. The nuclei of both trophoblast and embryoblast usually contain an elongated nucleolus [10] with a peripheral border of chromatin; sometimes, additional nucleoli are visible. Each cell invariably shows several chromocenters.

The total number of cells varies. Specimen KT 970/72 consisted of 27 embryonic and 98 trophoblastic cells. In addition, there is an indistinct polocyte adhered to the trophoblast.

The *zona pellucida* has completely disappeared. The *overall diameter* varies considerably because of deformation, shrinkage and other reasons [13]. For instance, after Carnoy's solution, 2 blastocysts measured 70×100 microns, a third 75×80 microns. The PAS-reaction shows

---

Figs. 23–29: Blastocyst, 101 h

Fig. 23	Ovary with 3 corpora lutea, low magnification. PAS. KT 967. 40:1
Fig. 24	Detail of Fig. 23. Corpus luteum with enlarged blood vessels. 115:1
Fig. 25	Uterine horn, longitudinal section. Blastocyst (*arrow*) antimesometrial. *M* Mesometrial region of uterus with enlarged blood vessels, PAS. KT 967. 40:1
Fig. 26	Blastocyst and vicinity, H.-E. *C* compact layer of endometrium beneath epithelium. Transverse ridges of the epithelium cause a wavy appearance of the surface. KT 970. 135:1
Fig. 27	Blastocyst of Fig. 26, enlarged. *Em* Embryoblast with mitosis. 720:1
Fig. 28	Blastocyst and vicinity, PAS. A crypt has formed. KT 967. 135:1
Fig. 29	Detail of next section (Fig. 28). Fine glycogen granules in trophoblast "*T*" and the adjoining epithelium "*U*". In between, a free space, evidently formed artificially by shrinkage (The blastocyst must originally have occupied the whole crypt). 560:1

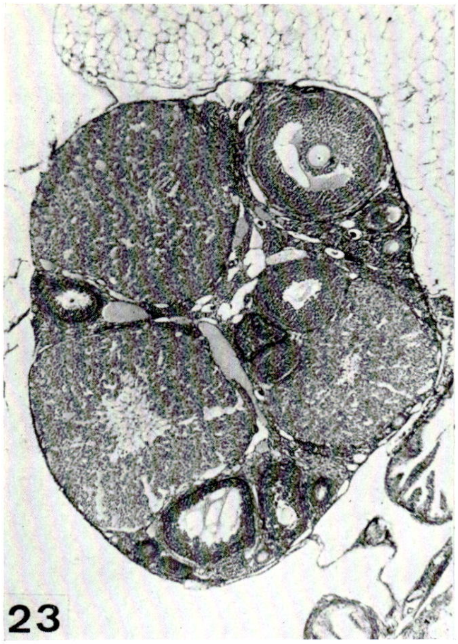
23

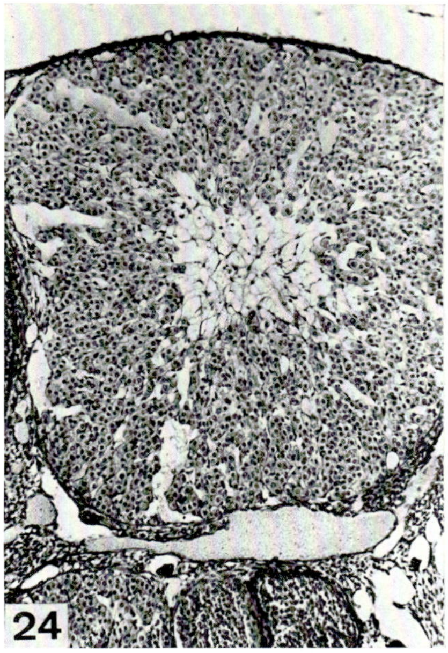
24

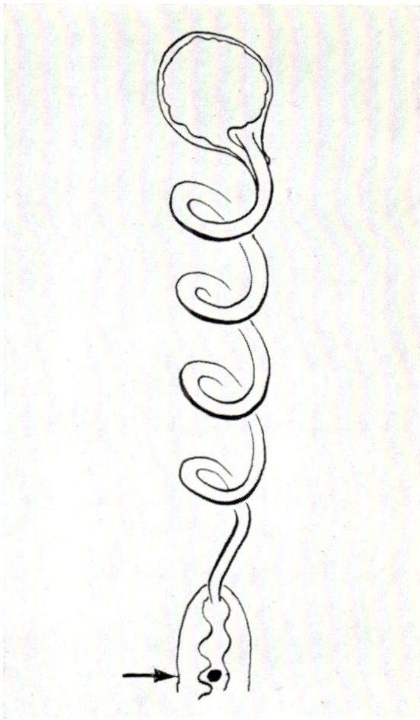

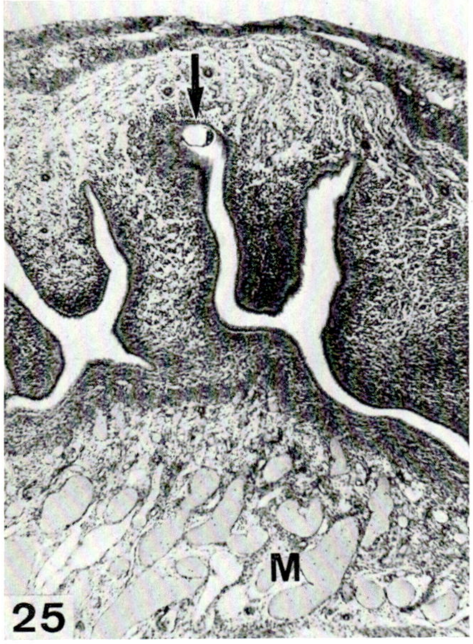
M
25

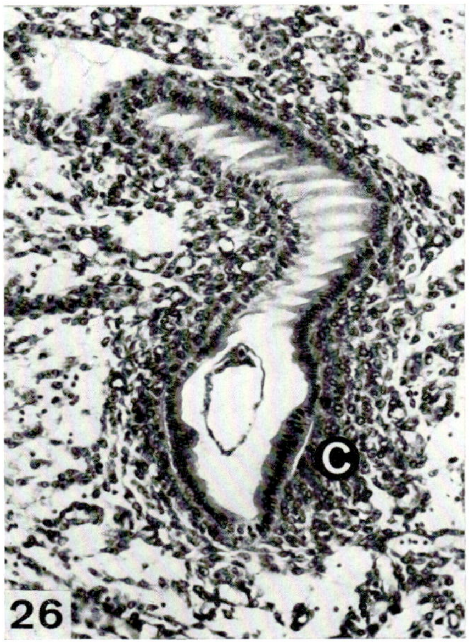
C
26

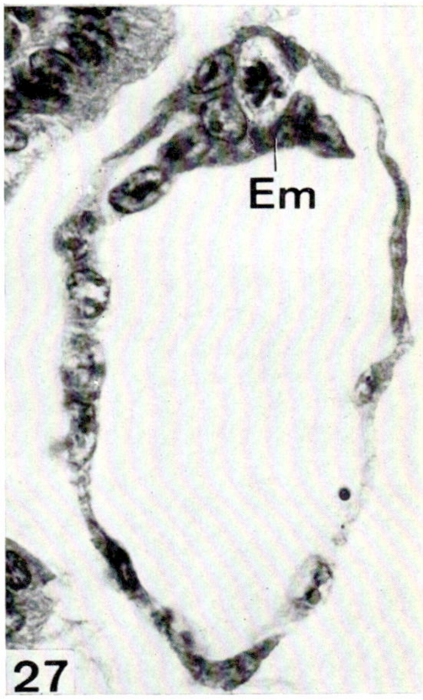
Em
27

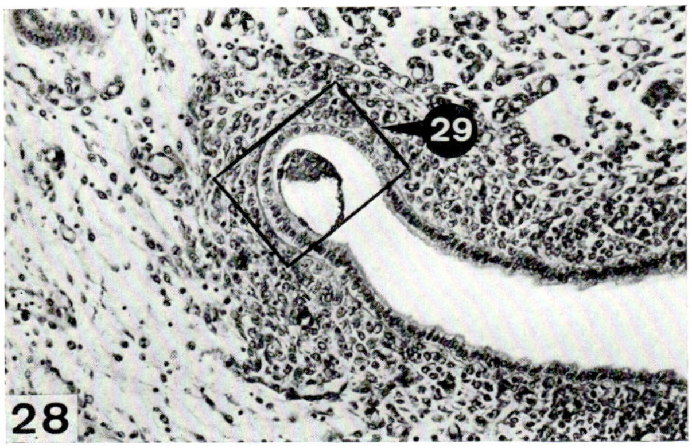
29
28

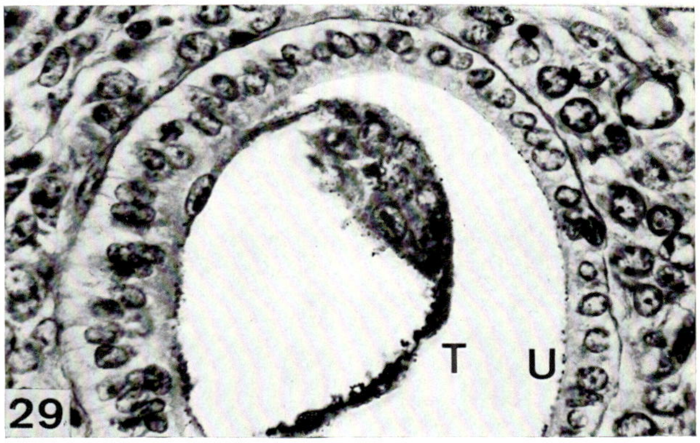
T
U
29

intense red granules within the trophoblastic cells after the zona pellucida disappears (Fig. 29). Some embryoblastic cells bordering the segmentation cavity may contain similar granules. Perhaps they represent the first entodermal cells.

Corpora lutea contain radially arranged strands of large clear cells, distinctly different from the smaller interstitial and follicle cells (Figs. 23–24). Blood vessels are abundant and sometimes dilated. Within the interstitium, some small clusters of PAS-positive cells with fine red granules may be observed (Fig. 37). They should not be confused with the intensely red debris of degenerating oocytes.

Material	Age	Blastocysts
KT 964	99 h	2 blastocysts in uterus, without zona pellucida
KT 969/70	101 h	3 blastocysts in left uterine horn 4 blastocysts in right oviduct
KT 967/68	101 h	4 blastocysts in uterus
KT 923	94 h	none visible, one empty zona pellucida

Stage 6 Implantation

4 1/2 Days

Horizon IV
Implantation

At 120 hours, there are distinct differences in degree of development of embryos within the same litter. Some embryos are *beginning implantation*, i.e. the blastocyst closely adheres to the undamaged uterine epithelium (Fig. 32). In other cases, there is advanced erosion of the mucous membrane. Some 4-day embryos are also closely adhered to the undamaged uterine epithelium, and so the beginning of implantation is estimated to occur at 4 1/2 days.
Embryoblast and trophoblast can easily be discriminated at this stage. Trophoblastic cells are flat, with prominent nuclei. Embryoblastic cells are spherical and have larger nucleoli than the trophoblastic cells. Both types of cells are rich in RNA. They stain intensively with Pyronin, in contrast to the underlying endometrium. The *entoderm cells* are already recognizable as a distinct layer and their cytoplasm appears even darker than other embryonic or trophoblastic cells. The number of cells varies. For example, specimen KT 859 (Fig. 31) consisted of 76 entodermal, 34 formative (embryoblastic) and 133 trophoblastic cells.

Invasion

Erosion of the uterine epithelium usually begins somewhat below the equatorial zone of the blastocyst. The adjoining trophoblastic cells are transformed into "*trophoblastic giant cells*" [2] (Fig. 35). The nuclei become large and spherical. The nucleoli also enlarge, and the cytoplasm forms long slender processes.
These cells are easily distinguished from the large *deciduous cells* which now appear in the vicinity of the implantation cavity [40]. They contain little RNA and much glycogen. In H.-E.-sections, the glycogen is dissolved and characteristic vacuoles are visible (Fig. 35). Adjacent to the zone of glycogen cells (Fig. 34, *D*) mentioned above, there is another girdle of cells consisting mainly of enlarged deciduous cells without glycogen droplets. Some of these cells have exceedingly large nuclei (Fig. 34, *D*). At the periphery, towards the muscle layer, the RNA-content of the endometrium decreases considerably.

Material	Age	Blastocysts
KT 859	117 h	1 free, 3 attached, all in one uterine horn. Entoderm distinct (H.-E.-stained)
KT 873	120 h	3 attached in uterus. Entoderm distinct (methylgreen-pyronin stained)
KT 857	124 h	1 free in uterine horn. Entoderm distinct (H.-E.)

Figs. 30–37: Implantation

Fig. 30 Low magnification of uterus, longitudinal section.
M Mesometrium (oblique section), H.-E.
KT 859. 117 h. 40:1

Fig. 31 Blastocyst, enlarged.
KT 859. 560:1

Fig. 32 Blastocyst, 120 h, Carnoy fixation, phase-contrast.
Little shrinkage.
KT 873. 225:1

Fig. 33 Endometrial reaction. Blastocyst attached, 117 h.
KT 859. 135:1

Fig. 34 Detail from Fig. 33.
G Girdle of glycogen cells, *D* zone of large deciduous cells. 360:1

Fig. 35 Invasion, high magnification.
The trophoblastic cell "*T*" is enlarged, and has penetrated the epithelium. In the connective tissue, glycogen containing cells with typical vacuoles "*V*" are visible, 117 h. 720:1

Fig. 36 Ovary with 4 corpora lutea, PAS, 117 h.
KT 911. 27:1

Fig. 37 Detail of ovary.
Left: margin of corpus luteum, bordering the interstitial gland with enlarged blood vessels and some intensely PAS-positive cells "*P*". 270:1

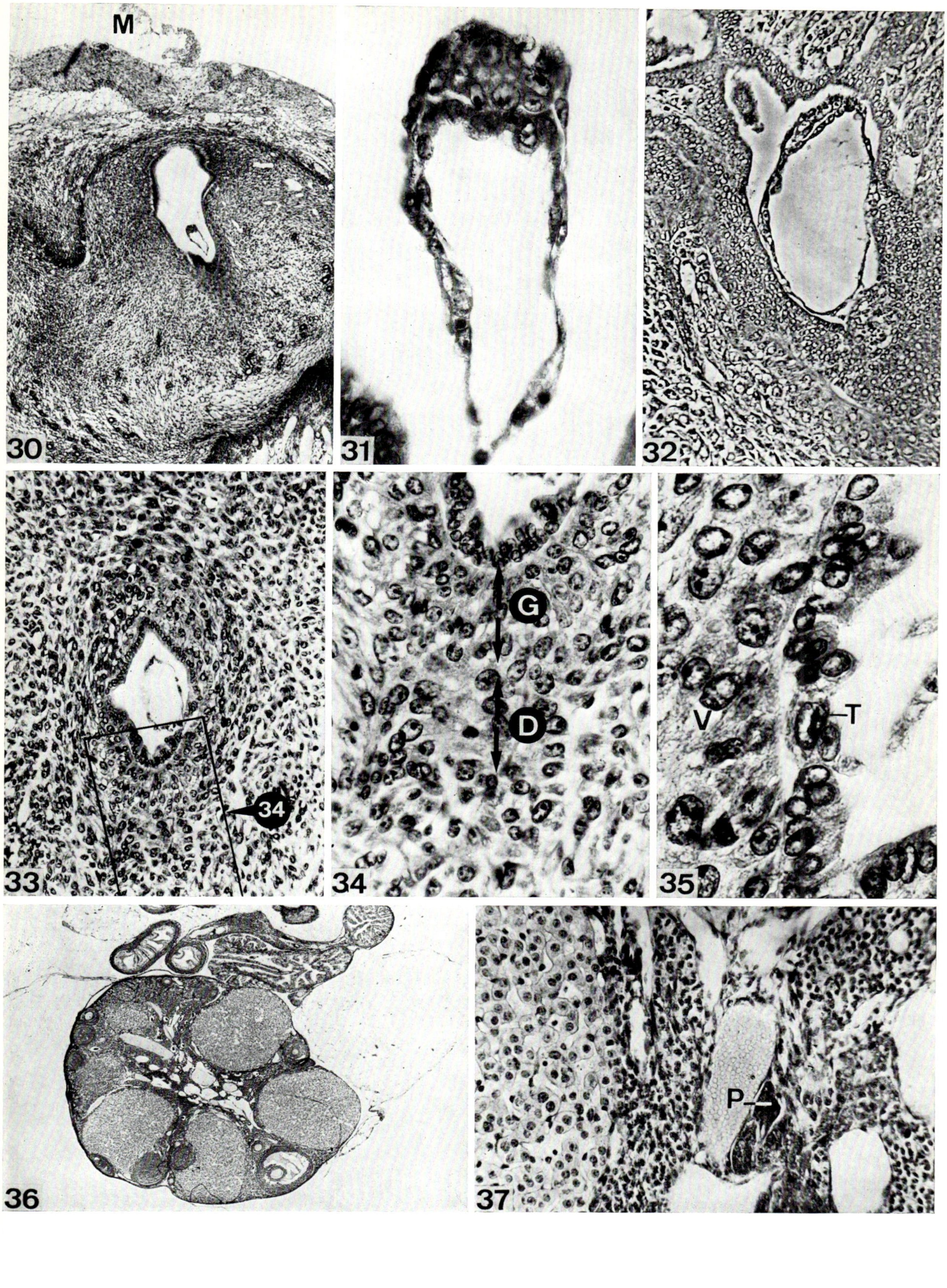
M
30
31
32
34
33
G
D
34
V
T
35
36
P
37

Horizon V
Ovum implanted, but still avillous

Stage 7 Formation of Egg Cylinder

5 Days

As soon as invasion starts, the embryoblast enlarges considerably and bulges cone-like into the segmentation cavity (Fig. 41). The trophoblast also starts to form an excrescence (trophoblastic cap or ectoplacental cone), which projects above the embryonic pole and stains intensively with H.-E. (Fig. 45).

Embryoblast

The formative cells (= "inner cell mass") of the embryoblast may be distinguished from the adjacent "trophoblastic cap" by their pale appearance. The detection of the entodermal cells, on the opposite side of the egg cylinder, is easier. They are cuboidal in shape and have been called proximal or visceral entoderm. Their surface does not appear to be sharply

---

Figs. 38–45: Formation of egg cylinder

Fig. 38 Low magnification of uterus, cross section, PAS, 117 h.
M Mesometrium.
KT 911. 40:1

Fig. 39 Detail: Deciduous reaction.
In the vicinity of the egg, the uterine epithelium is dissolved. Appearance of glycogen cells (dark border) and of large deciduous cells. Enlarged blood vessels, PAS, 117 h. 100:1

Fig. 40 Implantation site enlarged. The uterine epithelium has disappeared.
G Large dark glycogen droplets in glycogen cells, *T* trophoblastic giant cell with faintly staining nucleus and large nucleolus, PAS, 117 h. 560:1

Fig. 41 Low magnification of implantation site.
Thin section, OsO_4 fixation, phase-contrast.
KT 782. 360:1

Fig. 42 Detail of Fig. 41.
En Entoblast, with indistinct boundary (because of the presence of microvilli). 900:1

Fig. 43 Detail of Fig. 41.
E Large nucleus of embryonic cell, adjoining uterine epithelium. The latter contains lipoid droplets, sometimes confluent and forming spheres of variable size (*L*) stained black by OsO_4. 900:1

Fig. 44 Invasion and formation of egg cylinder, H.-E.
KT 693, 120 h. 100:1

Fig. 45 Detail of Fig. 44.
Cap-like thickening of trophoblast "*K*".
Arrows: limits of the uterine epithelium. 270:1

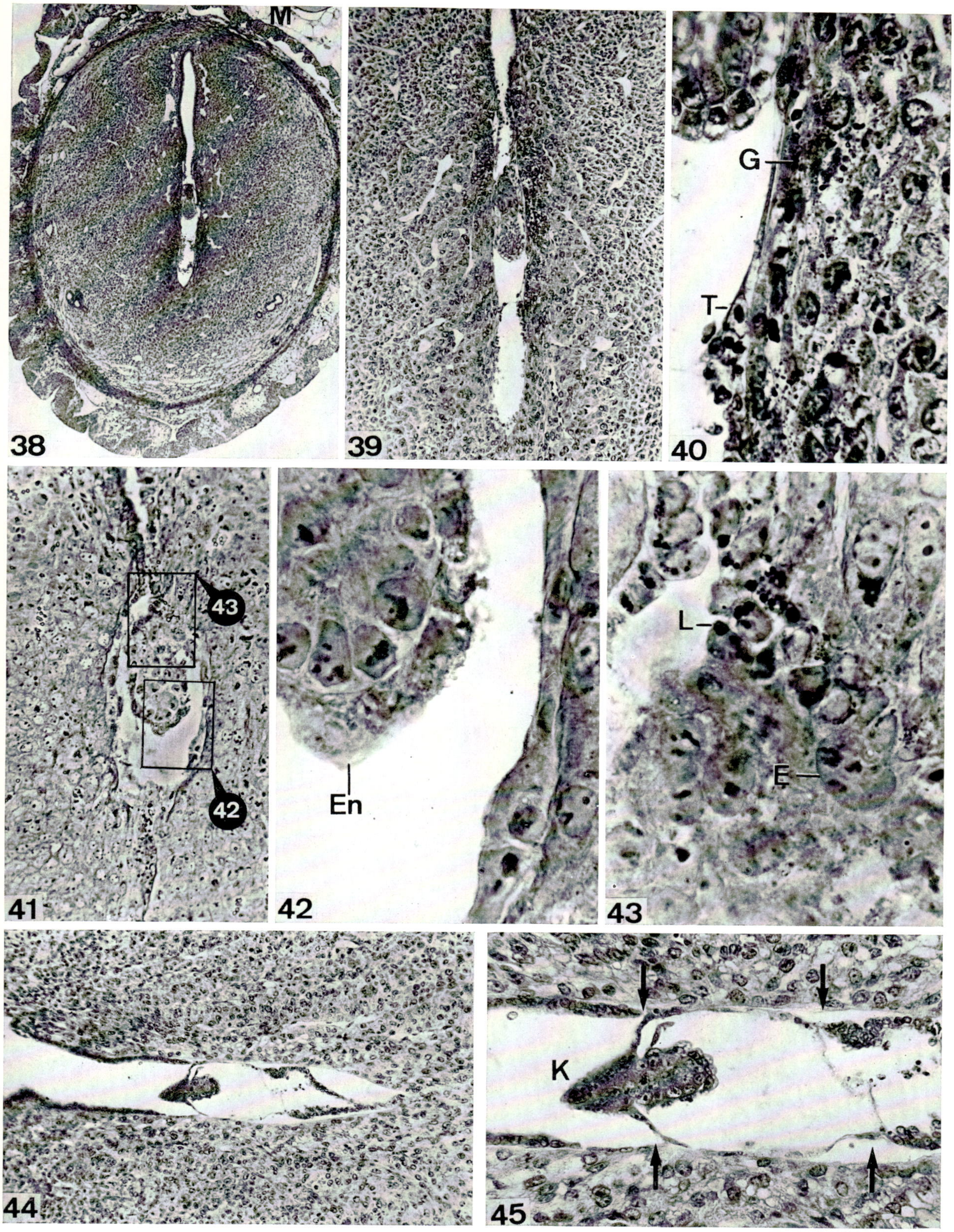
38
M
39
40
G
T
41
43
42
42
En
43
L
E
44
45
K

bounded with the light microscope because of the presence of microvilli (Fig. 42). Towards the base of the egg cylinder, near the trophoblast, the distinction between embryoblast and trophoblast is less obvious. Some cells migrate along the inner surface of the trophoblast at the base of the egg cylinder (Fig. 45) and become the distal or parietal entoderm.

Trophoblast

Above the egg cylinder the trophoblastic cells are cuboidal in shape and protrude as a cap-like mass (Fig. 45). Further growth of this trophoblastic cap gives rise to the "ectoplacental cone".

The peripheral, flattened trophoblastic cells dissolve the adjacent uterine epithelium. The disintegration proceeds all around the egg cylinder. Within the epithelial cells, large lipoid droplets are often visible (Fig. 43). They have been interpreted as "secretion droplets" [2].

Endometrial Reaction

The endometrial reaction involves both the uterine epithelium and connective tissue [22]. The adjoining epithelial cells degenerate and numerous lipoid droplets form within the cytoplasm. These are stained black with OsO_4 and red with Sudan B.

Fig. 44 shows some cell debris in the original uterine lumen (right). In the vicinity of the implantation site, the deciduous cells exhibit a marked glycogen reaction when stained with PAS ("deciduous glycogen cells").

Tubular uterine glands are now more numerous. They are situated in between implantation sites.

Material	Age	
KT 693	119 h	7 egg cylinders, stained with H.-E. only part of uterus fixed in OsO_4 for thin sections, in methacrylate
KT 782	117 h	2 egg cylinders
KT 911/12	117 h	6 egg cylinders, H.-E. and PAS

Horizon V
Ovum implanted,
but still avillous

Stage 8 Differentiation of Egg Cylinder

6 Days

Implantation sites are easily visible externally as spherical swellings, measuring 2×3 mm (Fig. 46). The spacing of the embryos is now rather regular [26]. However, in one instance, two embryos in one swelling were observed (specimen KT 914).

Embryoblast

The egg cylinder elongates and will soon be divided into *embryonic* and *extraembryonic* areas. In between the two areas there is a small indentation. Soon a central lumen, the *proamniotic cavity* appears, at first in the embryonic region, and later in the extraembryonic area. Most embryos of this stage have already formed a continuous cleft.

Most of the cells of the embryonic ectoderm are cylindrical in shape. They are often separated by a small furrow from the extraembryonic, irregular cuboidal cells.

The entodermal cells become flattened at the ventral end of the egg cylinder. Towards its base, they become cylindrical and accumulate much glycogen in their apical cytoplasm [29] (Fig. 51, lower left).

Trophoblast

Trophoblastic cells migrate irregularly from the *ectoplacental cone*, and maternal erythrocytes spill into the developing intercellular lacunae. Ectoplacental cells store much PAS-positive material, probably glycogen (Fig. 51). They are called by Arvis [23] "cellules ultratrophoblastiques". Their shape is more regular than the glycogen cells of the decidua.

The peripheral, previously flattened trophoblastic cells invade the maternal tissue and enlarge greatly. They are often called "*primary giant cells*" and contain a single, very large nucleus (Fig. 54). The peripheral cells of the ectoplacental cone also begin to enlarge.

Trophoblastic giant cells never undergo mitosis [30]. Their distinction from smaller, so-called "secondary giant cells" [2] (Fig. 107) is perhaps unwarranted as they are also derived from the trophoblast.

Reichert's Membrane

With the appearance of the distal (= parietal) entoderm, a non-cellular thin membrane, Reichert's membrane, forms between the distal entoderm and the peripheral trophoblast (= trophectoderm). Though acellular in nature, it increases in size with the growth of the embryo, and can be distinctly recognized during later development. It is secreted by the distal entoderm (see discussion in Green [2]).

Material	Age	Embryos	Lumen (proamnion)	Closure of mesometrial uterine lumen
KT 913/14	5 days 23 h	(left) 2	+	halfway, hemorrhage
		(right) 3	+	halfway, hemorrhage
KT 926/27	5 days 23 h	(left) 3	none	no
		(right) 1	none	no
KT 928/29	5 days 23 h	(left) 1	+	halfway, hemorrhage
		(right) 3	+	halfway, hemorrhage
KT 723	6 days 3 h	total 6	+	halfway, hemorrhage
KT 919	6 days 3 h	(left) 4	+	more than half, hemorrhage
		(right) 4 two of them in resorption	+	more than half, hemorrhage

Figs. 46–51: Differentiation of egg cylinder: 5 days 23 h

Fig. 46 Drawing of uterus at 5 days 23 h.
KT 928/29

Fig. 47 Low magnification of uterus, cross section.
M Mesometrium, *U* disappearing uterine lumen.
KT 914. 34:1

Fig. 48 Implantation site. The ectoplacental cone appears diffuse in its peripheral region.
Z Decaying uterine epithelium, in the original lumen. 100:1

Fig. 49 Egg cylinder, retarded in development by about 24 h.
KT 926, 5 days 23 h. 270:1

Fig. 50 Differentiation of egg cylinder.
KT 929, 5 days 23 h. 100:1

Fig. 51 Detail of Fig. 48.
Some scattered cells of the ectoplacental cone have differentiated into ectoplacental glycogen cells (*eG*, dashed boundaries). Maternal blood (*B*), only faintly stained, fills the interspace. 270:1

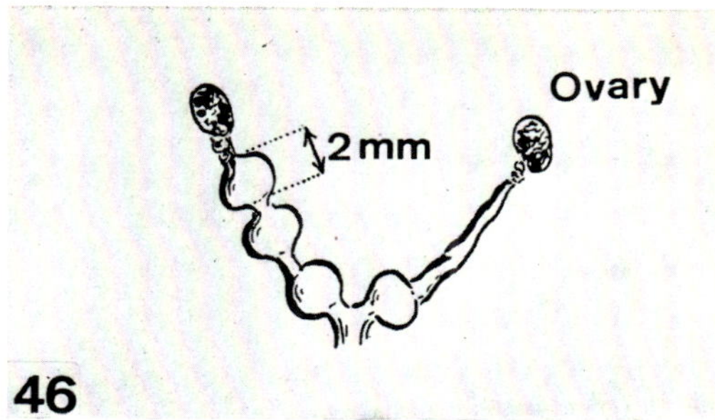

46

47

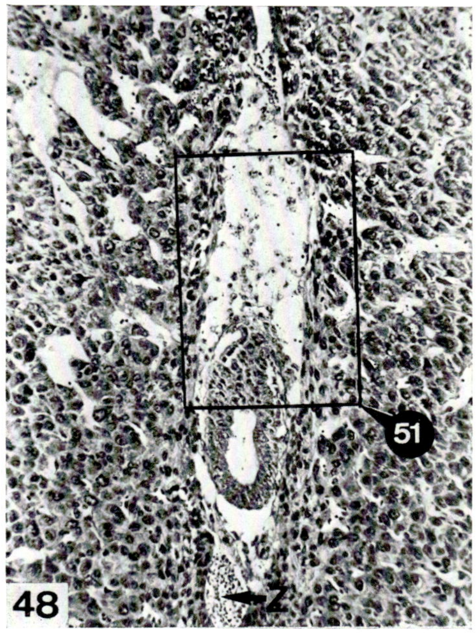

48

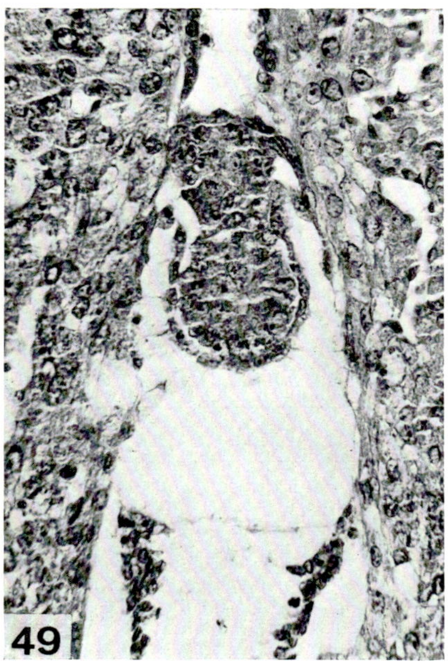
49

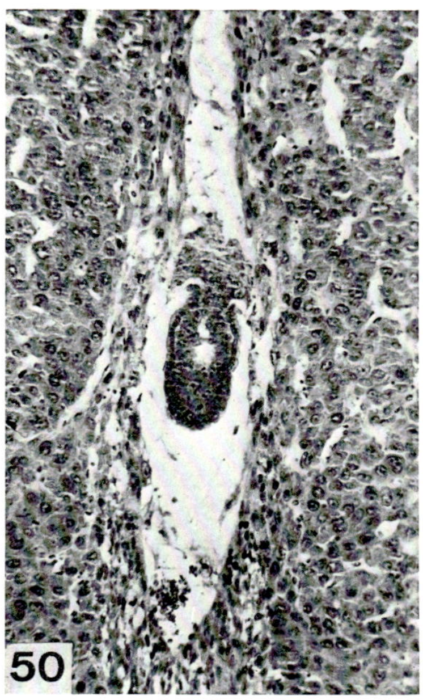
50

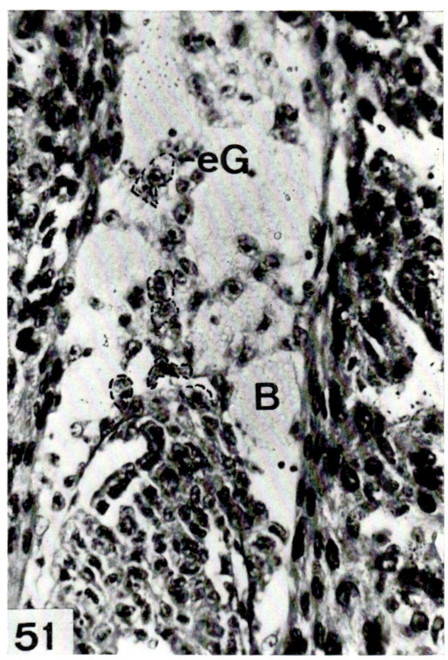

51

Horizon VI
Primitive villi,
distinct yolk sac.
Horizon VII
Branching villi,
axis of germ disc defined

Stage 9 Advanced Endometrial Reaction

6 1/2 Days

By 6 1/2 days, the structure of the uterine mucosa is not much different from what it was on the 6th day. By 7 days, it still may not have changed, mainly because of the variation in the degree of development of specimens having the same copulation age. The uterine reaction was examined in detail in a specimen of 6 days 20 hours which was somewhat retarded in development. The specimen is regarded as typical of 6 1/2 days. The structure of the implantation site, rather than the structure of the embryo, was chosen by STREETER [6] to define the comparable human age.

Uterine Reaction

After the disintegration of the uterine epithelium, the ectoplacental cone is invaded by blood (Fig. 51). The original lumen of the uterine crypt has disappeared. In this way, the developing placenta gains a solid contact with its environment and with the mesometrial blood vessels. The *deciduous cells* are manifold in appearance (Fig. 55). Deciduous glycogen cells with coarse droplets are seen near the periphery of the deciduous transformation zone, predominantly in the mesometrial direction. At the same time, large capillaries invade this area. Probably the capillary invasion is necessary for the rapid production of glycogen. The other deciduous cells contain only fine, diffusely distributed glycogen. Many mitoses can be found, sometimes irregular in appearance, and they resemble those in malignant tumors. This impression is enhanced by the appearance of several multinucleated giant cells (Fig. 55).

Embryonic Axis

During this stage of development, the embryonic axis is determined. A limiting furrow *F* (Fig. 56) is situated cranially, at the front end of the embryo. Later on it deepens (Fig. 66). In the mouse, no clearly prominent primitive knot appears, and a defined primitive groove does not form until later.

Material	Age	
KT 654/55	6 days 20 h	7 egg cylinders, H.-E.

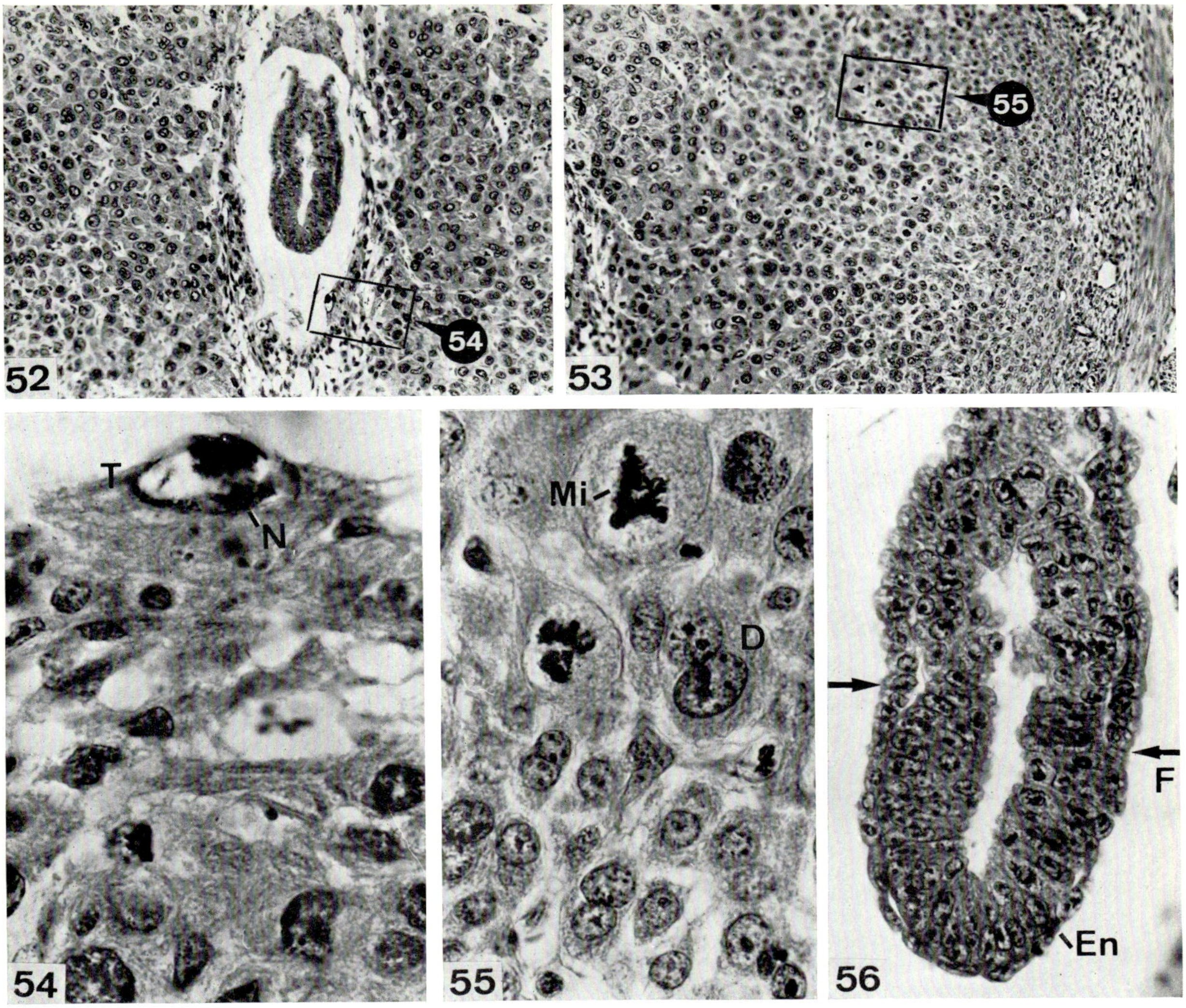

Figs. 52–56: Advanced endometrial reaction: 6 days 20 h

Fig. 52 Implantation site, low magnification.
An artificial cleft is seen between egg cylinder and ectoplacental cone.
KT 654. 100:1

Fig. 53 Deciduous boundary zone, low magnification.
KT 654. 100:1

Fig. 54 Detail of Fig. 52.
T Primary, trophoblastic giant cell with clear spherical nucleus containing a large nucleolus (*N*) and coarse chromatin which is partially attached to the nucleolus. 700:1

Fig. 55 Detail of Fig. 53.
Boundary zone of deciduous expansion.
Irregular mitosis (*Mi*) and multinucleate deciduous cells (*D*). 700:1

Fig. 56 Egg cylinder with proamniotic cavity.
The boundary between embryonic and extraembryonic area is marked with *arrows*. "*F*" is the cranial limiting furrow of the embryo. The entoblastic cells become flattened at the free end of egg cylinder (*En*).
KT 654. 270:1

Horizon VIII
Hensen's node,
primitive groove

Stage 10 Amnion

7 Days

At 7 days, the shape of the extraembryonic part of the egg cylinder changes rapidly. A crescent-like *transverse fold* [41] temporarily appears, and pushes into the proamniotic cavity just caudal to the primitive streak. It was observed as a regularly occurring structure by SNELL and STEVENS [2]. It is perhaps a result of the rapid growth that occurs in the posterior wall of the egg cylinder at this time. It disappears in the following stage.

Formation of Amnion

The tissue at the posterior end of the primitive streak bulges into the proamniotic cavity, and forms the *posterior amniotic fold*. It is continuous with the smaller lateral amniotic folds which unite to form the anterior amniotic fold. In this way, a continuous constriction forms around the middle of the egg cylinder, which is drawn tighter and tighter as the folds develop (Fig. 64). Finally, the lips of the folds will fuse and the amniotic cavity will be sealed off completely. In all embryos of this stage it is still open.
In the mesoderm of the posterior amniotic fold, small cavities appear between the cells, which coalesce to form a single large cavity, the *exocoelom*. The exocoelom is lined, except for the allantois, by a mesothelium.

Figs. 57–63: Formation of amnion: 7 days

Fig. 57 Low magnification of uterus, cross section.
U Remnant of uterine lumen, specimen bt 76, 7 days.
No extraembryonic transverse fold. 22.5:1

Fig. 58 Low magnification of uterus, cross section.
Zo Vascular zone developing. The section passes closely lateral to the connecting stalk of the egg cylinder.
KT 948, 7 days 3 h. 25:1

Fig. 59 Reconstruction, starting from a parasagittal section.
Dashed line: hidden lumen of egg cylinder, *K* ectoplacental cone, *Q* transverse fold, below: the bulge of the amniotic fold.
KT 948, 7 days 3 h

Fig. 60 Egg cylinder.
KT 948, 7 days 3 h. 130:1

Fig. 61 Detail of Fig. 60.
C Lumen formation in mesoderm of the posterior amniotic fold, *Q* extraembryonic transverse fold. 350:1

Fig. 62 Sagittal section through egg cylinder.
bt 76, 7 days. 90:1

Fig. 63 Detail of Fig. 62.
En Entoderm, *Ek* ectoderm, M mesoderm. 270:1

57

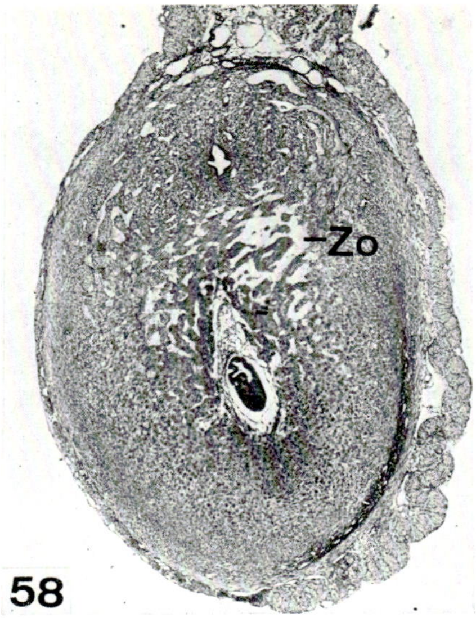

58

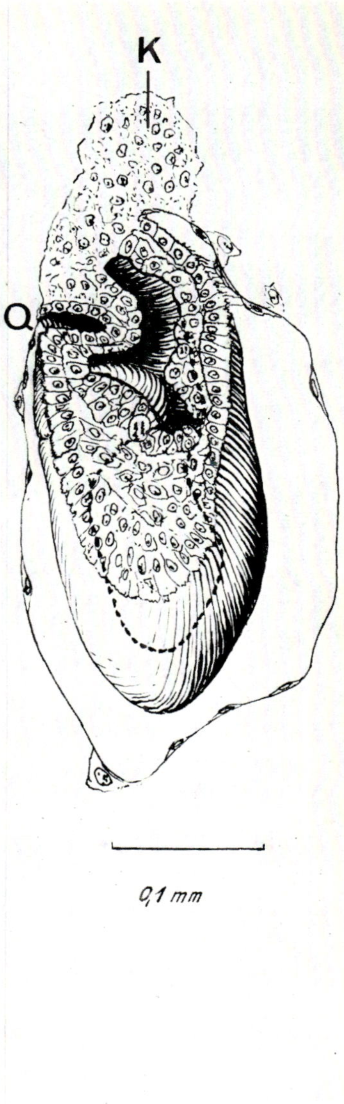

59

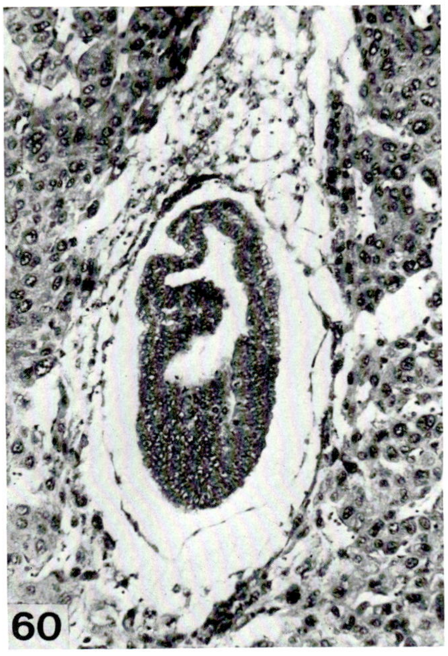
60

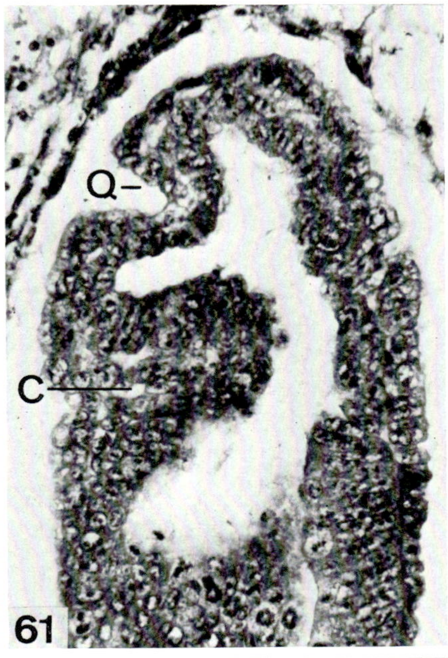

61

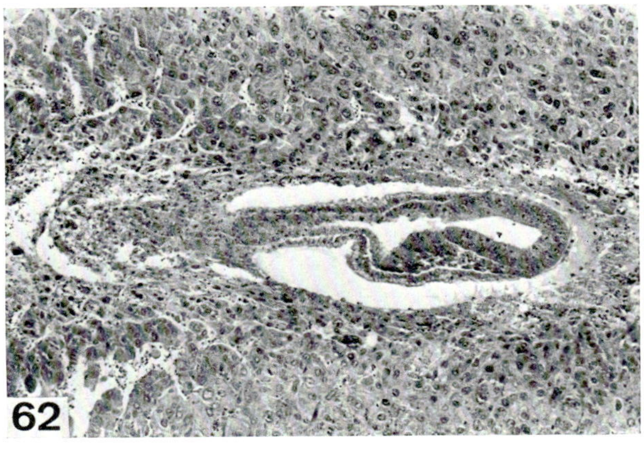
62

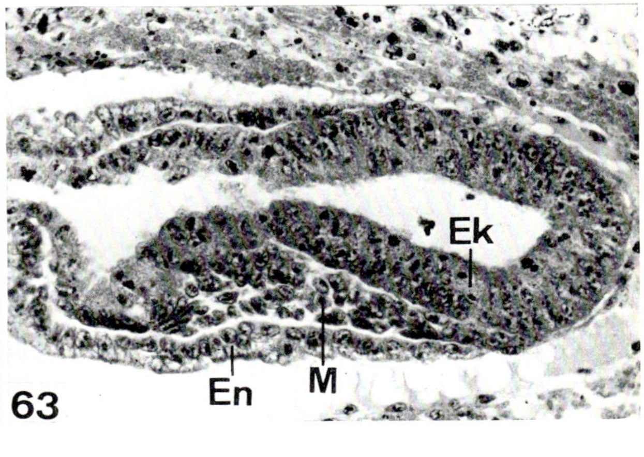

63

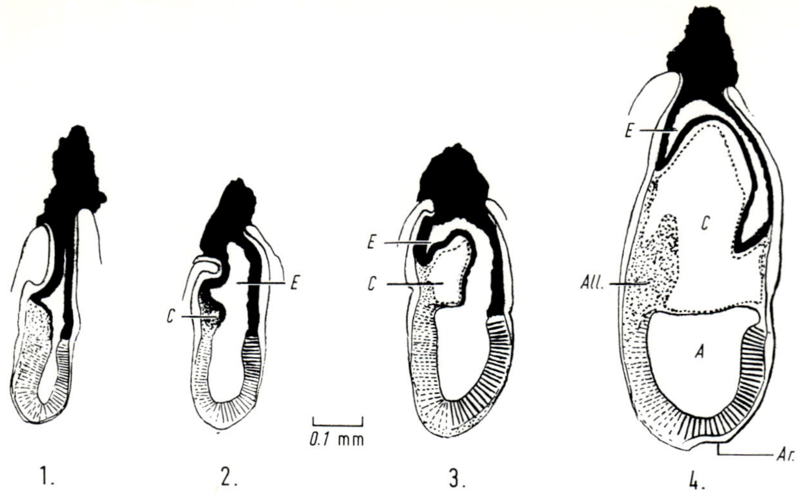

Fig. 64
Formation of amnion. 1. 7 days, 2. 7 days 3 h, 3. 7 days 10 h, 4. 7 days 20 h. The extra-embryonic coelom is developing (3rd Fig.) by enlargement of the tiny lumina *C* (2nd Fig.). The coelomic mesothelium (dashed line) pushes between the amniotic cavity "*A*" and the ectoplacental cavity "*E*". The allantois (*All*) at first grows free into the coelom. Extraembryonic ectoderm: black. *Ar* Archenteron

Formation of Mesoderm

The first mesoderm cells appear at the posterior end of the egg cylinder. As a consequence, its wall thickens and becomes three-layered (Fig. 63). In sagittal sections, the wedge-shaped head process may be observed between the flattened entoderm (*En*) and the ectoderm (*Ek*). The notochord cannot be yet identified.

Entoderm

Towards the ectoplacental cone, the cells of the visceral entoderm are cylindrical in shape and possess many vacuoles. These do not contain lipids or glycogen. However, in cryostatic sections very fine lipid droplets may be seen in the other parts of the cytoplasm. After fixation in Carnoy's solution, there is an intense PAS-reaction which is strictly limited to the superficial layer. The stained polysaccharide does not seem to be glycogen [29].

Endometrium

In the mesometrial side of the uterus, the capillaries enlarge and form distinct sinusoids (Fig. 58). The original *lumen of the uterus* is adjacent to this area, and is sometimes narrowed to a small continuous tube. The lumen becomes discontinuous in the next stage.

Material	Age	
KT 889	7 days 0 h	1 egg cylinder with extraembryonic transverse fold
KT 947/48	7 days 3 h	8 egg cylinders with extraembryonic transverse fold 1 young egg cylinder without extraembryonic transverse fold, without mesoderm
Bt 76	7 days	1 egg cylinder without extraembryonic transverse fold, but formation of mesoderm visible

Horizon IX
Neural folds,
elongated notochord

Stage 11 Neural Plate, Presomite Stage

7 1/2 Days

Some retarded specimens of 7 days 20 hours were included in this stage. Developmental differences between embryos of the same nominal age still exist.

Closure of Amnion

At 7 1/2 days, the amniotic cavity is sealed off. From now on, there are three separate cavities: amniotic cavity, exocoelom and ectoplacental cleft (Figs. 64 and 72b). For awhile the cleft extends as the *ectoplacental duct* to the amnion. Shortly before the development of the foregut, this duct is pushed away from the front wall of the egg cylinder by the enlarging exocoelom. It persists for awhile as a short blind extension.

The neural plate is clearly delimited anteriorly and laterally. In the midline, it forms a shallow groove. Posteriorly it is less clearly defined and merges with the primitive streak (Fig. 70). In front of the neural plate, a small oral plate consisting of two epithelia can be seen. It is just posterior to the small heart rudiment (Fig. 74).

The head process is now developing. The flattened entodermal cells situated at the free end of the egg cylinder disappear. Therefore, the head process is directly exposed to the lumen of the yolk sac. It is obviously intercalated secondarily into the entoderm and cannot be distinguished from it with certainty. In the midline, its cells become cylindrical; posteriorly they form the *notochordal plate* which is slightly indented. This indentation has also been called the *archenteron* (Fig. 69).

The archenteron is a transitory structure which has nothing to do with the formation of the hind gut. The development of the notochordal plate is similar to the formation of the notochordal canal in the human. In mice, no canal and no distinct primitive pit exist. A "canalis neurentericus" is also completely lacking.

Foregut

At 7 days, a small furrow appears in the entoderm beneath the anterior amniotic fold (Figs. 66 and 67, "*F*"). It is situated at the boundary between the embryonic and extraembryonic region, *anterior* to the developing heart rudiment, in the area of what later will be the septum transversum [41]. The structure of the entodermal cells in this area changes markedly. Sometimes this change can already be seen at 6 days (in Fig. 50 represented by a slight constriction in the middle of the right wall of the egg cylinder). The foregut pocket will develop *caudal* to this constriction (Fig. 74).

Material	Age	
KT 996	7 days 10 h	8 egg cylinders. Formation of amnion 1 pathologic egg cylinder, without lumen
KT 954/55	7 days 20 h	8 neurulae, presomite stages 2 of them implanted in close proximity
KT 687	8 days 0 h	4 neurulae, presomite stages 2 somite stages: 1 and 2 somites; deep foregut portal 1 in resorption

Figs. 65–71: Neural plate, presomite stage, $7^1/_2$ days

Fig. 65 Low magnification of uterus, cross section.
U Remnant of uterine lumen.
KT 996/2, 7 days 10 h. 24:1

Fig. 66 Detail of Fig. 65.
H Posterior amniotic fold, *F* furrow in entoderm. The exocoelom (marked throughout with *) is also visible within the anterior amniotic fold.
KT 996/2, 7 days 10 h. 270:1

Fig. 67 Tangential section of lateral amniotic fold, with exocoelom*.
Q Extraembryonic transverse fold, *E* ectoplacental cavity, *F* fold in entoderm.
KT 996/3, 7 days 10 h. 130:1

Fig. 68 Low magnification, slightly older stage.
M Mesometrium.
KT 954/1, 7 days 20 h. 20:1

Fig. 69 Head process with Archenteron "*Ar*".
A Amniotic cavity, *B* blood islet, *E* ectoplacental cavity, * = exocoelom, *K* head fold.
KT 954/1, 7 days 20 h. 100:1

Fig. 70 Detail of Fig. 68
showing allantois (*All*), *E* ectoplacental cavity, *A* amniotic cavity, * = exocoelom, *P* primitive streak.
KT 954/1, 7 days 20 h. 100:1

Fig. 71 Higher magnification of allantois (*All*) and Reichert's membrane (*R*).
E Ectoplacental cavity, * = exocoelom.
KT 954/1, 7 days 20 h. 270:1

U
65

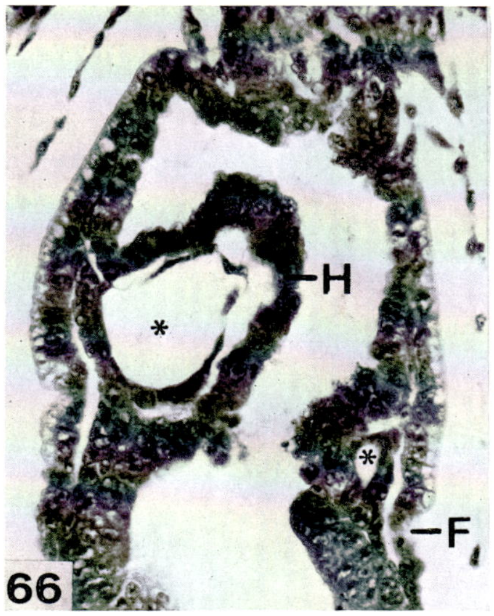
H
*
*
F
66

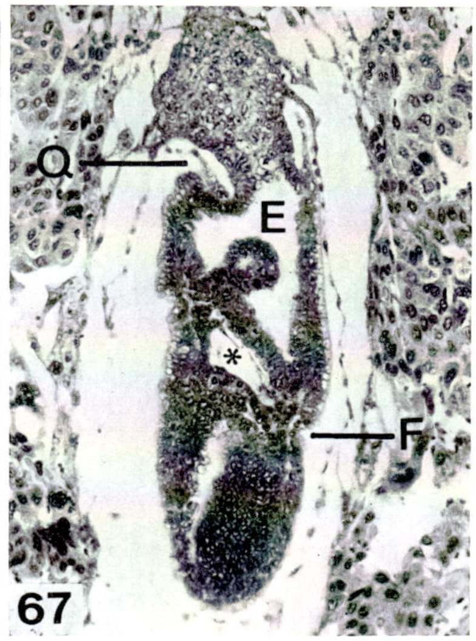
Q
E
*
F
67

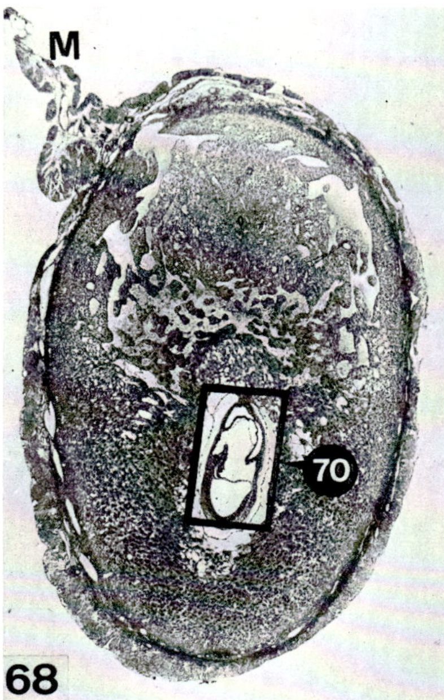
M
70
68

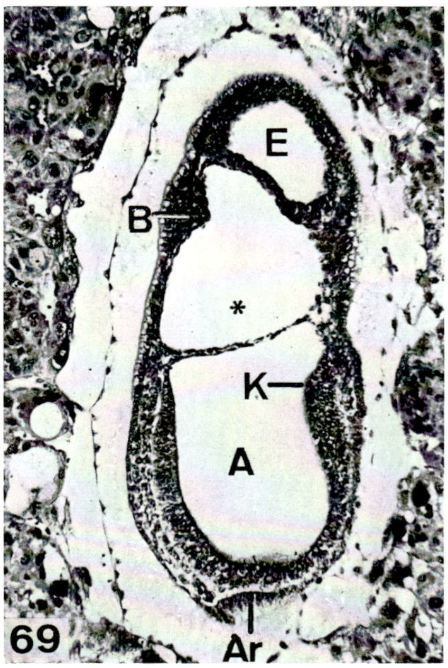
E
B
*
K
A
Ar
69

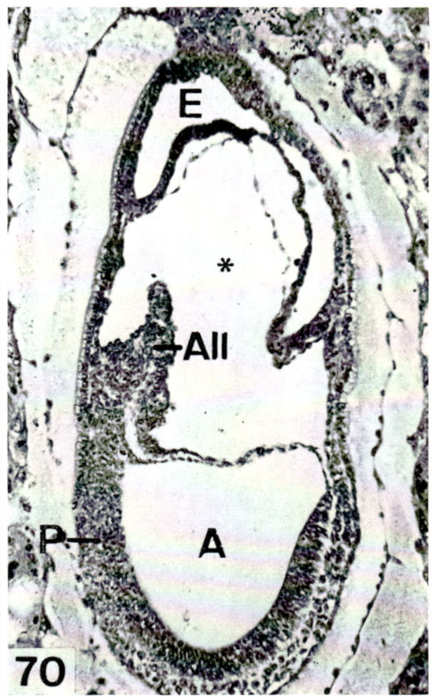
E
*
All
P
A
70

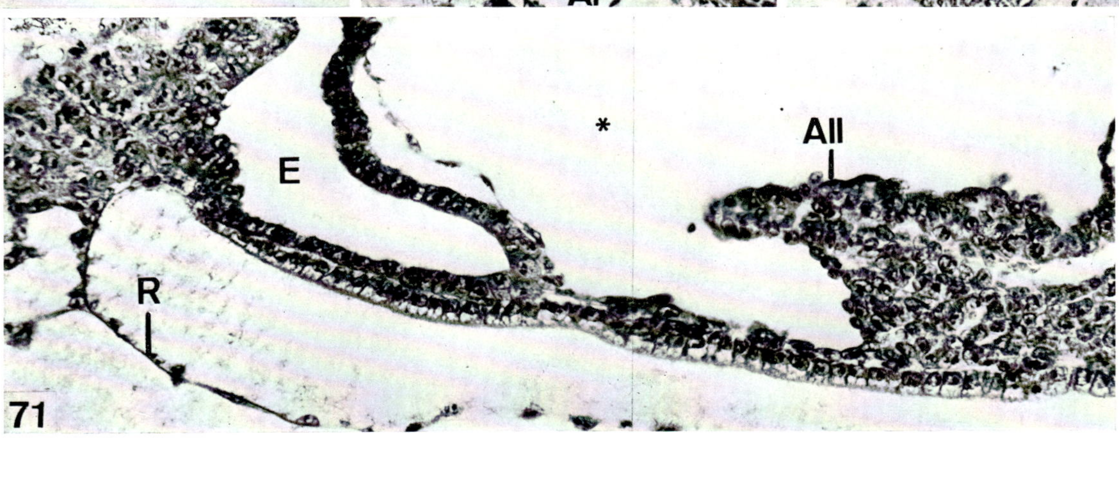
E
*
All
R
71

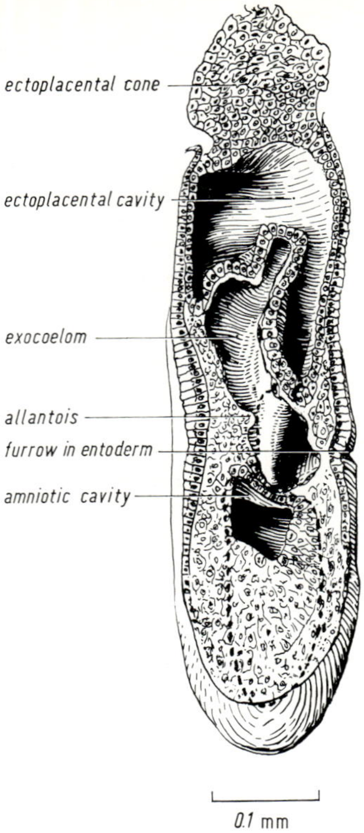

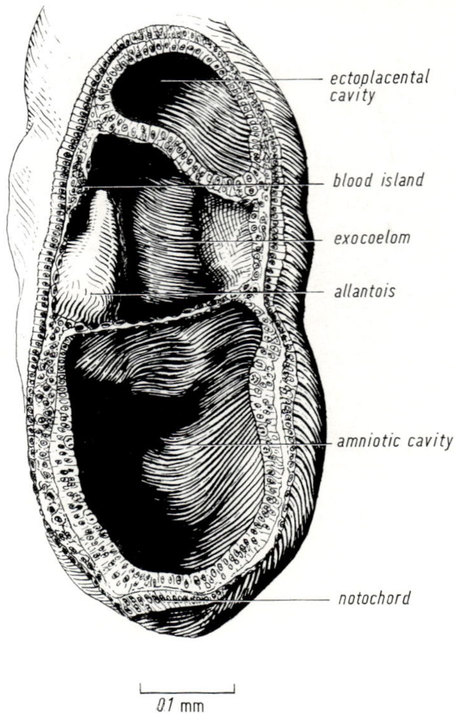

Fig. 72a
Cut egg cylinder, shortly before closure of amnion.
The section passes obliquely to the sagittal plane, through the right amniotic fold. Dashed line: contour of the amniotic cavity and connection with the ectoplacental cavity (ectoplacental duct).
KT 996/4, 7 days 10 h

Fig. 72b
Cut egg cylinder of 7 days, 20 h (after closure of amnion)

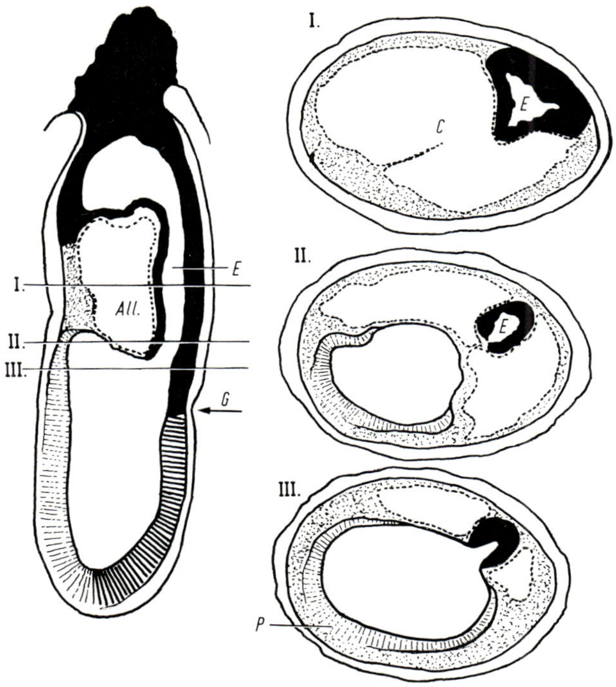

Fig. 73
Ectoplacental duct, shortly before closure of the amnion.
The drawing (left) shows the location of cross sections *I–III*.
E Ectoplacental duct, *P* transition of primitive streak and extraembryonic mesoderm, *C* mesothelial septum (in regression), *All* allantois, *G* limiting furrow

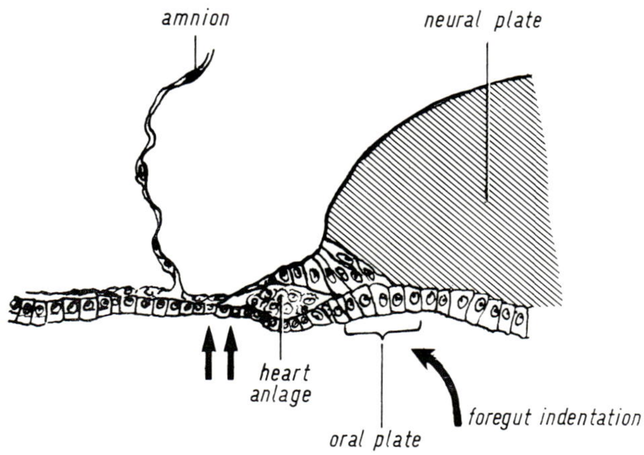

Fig. 74
Development of the foregut pocket.
Double arrow marks the embryonic–extraembryonic boundary. Here the epithelium is markedly reduced in height.
KT 955/5, 7 days 20 h

Horizon X/a
Younger group
Early somites present

Stage 12 First Somites

8 Days, 1–7 Somites

The copulation age of this group ranges from 7 days 21 hours to 8 days 21 hours. The seven somite embryo was chosen as the most advanced in this stage because with the formation of the eighth somite, the embryo begins to rotate around its longitudinal axis. In the human, the corresponding age group, as described by Streeter [6], extends from 1 to 12 somites. The number of somites is the most reliable criterion to determine developmental age.

External Form

A conspicuous characteristic of this stage is the progressive deepening of the neural groove. There is a marked dorsal (lordotic) flexure, and rotation has not yet begun. The head fold will bulge within a few hours (Figs. 76 and 80). The *anterior intestinal portal* deepens simultaneously. The posterior intestinal portal is beginning to form. It should not be confused with the archenteron (Fig. 77).

The *allantois* has grown out far into the exocoelom toward the ectoplacental cone. In a specimen of 6 somites it was still unattached, whereas in a littermate of 7 somites (KT 639/b 4) it had made firm contact with the developing chorion. Fig. 77 shows a contact already established in a 5-somite embryo.

The allantois is continuous cranially with the *primitive streak*. The latter is limited anteriorly by a "quasi primitive knot", which is caused by the bulging of the underlying archenteron (Figs. 77 and 80). If the archenteron forms only a shallow groove, this prominence is not distinct (Fig. 82).

Circulatory System

Blood islets develop during the preceding stage (7 1/2 days) in the wall of the yolk sac (Fig. 69). They are arranged as a girdle encircling the exocoelom, and may be seen in each section. Fig. 83 shows the marked mitotic activity of the hemocytoblasts [46]. There is no vascular connection yet with the vessels developing in situ in the body of the embryo. The *A. vitellina* arises first as paired vascular islands in the wall of the posterior intestinal portal (Fig. 82). Later, these anlagen unite to form a single, unpaired vessel [49].

The *heart rudiment* develops rapidly as seen, for instance, in 6-somite embryos. The first visible sign is a thickening of the mesoderm surrounding, horseshoe-like, the front end of the embryo. This cellular strand can even be seen in sections of presomite stages (Fig. 74).

The pericardial cavity first develops as lateral clefts in the mesoderm. At the 2-somite stage these intercellular gaps are clearly visible on each side of the midline (Fig. 81). At the 6-somite stage a distinct lumen also appears in median sections (Fig. 82). At the same time, single mesodermal cells have joined to form an endocardial tube. This tube is continuous with the *first aortic arch* which developed in situ along with the paired dorsal aortae.

The dorsal aorta originates bilaterally in the trunk region, beneath the somite stalks, and is visible in many cross sections (Fig. 79).

Intestinal Tract

The anlage of the foregut pouch has appeared in the preceding age group (Fig. 74). The epithelium of the gut is columnar in this area and adheres, for a short distance, to the high columnar epithelium in front of the neural plate. Both epithelial layers together form the *oral plate*. With the appearance of the first somite, the shallow groove quickly deepens to form a curved pocket (Fig. 82) narrowed ventrally by the prominent cardiac bulge. Most of the gut epithelium covering the bulge is cuboidal. The rudiment of the thyroid gland and of the liver develop here, but they cannot be distinguished yet with certainty. At the end of this period, the thyroid anlage appears as a distinct thick epithelial plate.

The notochord is situated in the dorsal wall of the foregut pocket. It is still intercalated in the entoderm. Later it will become separated. It consists of a long strand of columnar cells (Fig. 78) and still has a shallow groove posteriorly (the archenteron). The gut epithelium is flattened where it joins the notochord. Anteriorly, the wide first branchial cleft forms in the foregut pocket.

At the 2-somite stage the *posterior intestinal portal* is visible as a slight depression (Fig. 80) which deepens more slowly than the anterior one. At the 6 somite stage, it is only a slight depression (Fig. 82). The lining entodermal cells are higher than other cells in the area. Further caudally, in the extraembryonic area, is the epithelium of the visceral yolk sac with its characteristic cylindrical cells with vacuolated cytoplasm and basally located nuclei (Fig. 83). The superficial layer of the cytoplasm is strongly PAS-positive. The *cloacal membrane*, in contrast to the comparable stage in human embryos, has not yet appeared.

The *coelom* originates in the trunk region within the thickened marginal band of mesoblastic cells. Vesicles appear between the mesoblastic cells which coalesce to form a continuous mesoblastic cavity (Fig. 79). It is in open communication with the pericardiac cavity. In the older members of this group, it opens freely into the exocoelom on both sides.

Central Nervous System

In this stage, the brain plate develops very rapidly, and is the chief determiner of the embryonic form (Fig. 75). At 7 somites, the neural folds close at the level of the 4th and 5th somite, i.e., at the cervico-cranial boundary. From here, the closure proceeds both in anterior and posterior direction, in a zipper-like manner. Near the still widely open cranial end, a bilateral depression may be seen, the *optic evagination* (sulcus opticus, Fig. 75). Further caudally, two *neuromeres*[162] are visible. They are called "rhombomere *A* and *B*", homologous to the neuromeres in the human (Fig. 82). Rhombomere *B* is found at the level of the *otic placode*, which is visible at 4 or 5 somites.

In the area of the *cranial neural crest* [161] the trigeminal and facial crest are distinct.

Extraembryonic Membranes

In the preceding stage, three separate cavities were formed: amniotic cavity, exocoelom and ectoplacental cavity (Fig. 64). During this period the exocoel and the amniotic cavity expand at the expense of the ectoplacental cavity. Towards the end of this period, both ectoplacental layers fuse in the middle (Fig. 84). Laterally, both layers (*laminae*) are still recognizable (Fig. 77). At the same time, the *allantois* grows rapidly across the exocoelom, towards the ectoplacental cone. The ectoplacental cells are delimited from the exocoelom only by a thin mesodermal lining. The first contact of the allantois is made with these cells which frequently, at the point of contact, are detached from the adjacent ectoplacental tissue. It appears as a bulge toward the approaching allantois (Fig. 80). A more laterally situated point of contact occurs rarely (for example see Fig. 77). The time of fusion appears to vary also. It does not occur before the 5 somite stage, and is usually observed at 7–8 somites. The allantois itself transforms in its distal part into a loose meshwork of cells. Between them, endothelial-lined cavities develop which are the forerunners of the allantoic vessels (= umbilical artery and vein).

The *yolk sac* is composed primarily of two layers, the thin parietal (= distal) and the thick visceral (= proximal) entoderm. As described previously, only the vascular area of the visceral layer develops into the yolk sac proper. It enlarges now considerably. In later stages, this membrane can easily be recognized when the uterus is dissected. It is the inner lining of the narrow yolk sac cavity. Externally, the cavity is bound by the thin parietal layer and *Reichert's membrane* (Fig. 71). Adherent to Reichert's membrane is a layer of *trophoblastic giant cells* which grows rapidly in thickness, and finally forms a loose network of cells (Fig. 84, indicated by small circles). This area attaches the embryonic to the maternal tissue.

The adjoining *endometrium* has some multinucleate giant cells, already seen in the previous stage (Fig. 55). The decidua capsularis contains only a few blood vessels, whereas the decidua basalis has large sinusoids.

In the *ectoplacental cone* a special zone of glycogen-containing cells [29] becomes visible after PAS-staining (Fig. 84, "*eG*"). The cells are also seen in the preceding stage, but they were less numerous (Fig. 51, 6 days). With H.-E.-staining, they are inconspicuous, however they can be recognized by their numerous vacuoles in the faintly staining cytoplasm. Abundant storage of glycogen is characteristic of the cone cells which are differentiating into tropho-spongium (= junctional zone) [38].

Figs. 75–81: First somites, 8 days

Fig. 75 Dorsal view of 7-somite embryo.
Arrow marks the optic sulcus, *So* somite 5. 63:1

Fig. 76 Low magnification of uterus, cross section.
M Mesometrium.
KT 957/4, 7 days 21 h, 5 somites. 18:1

Fig. 77 Detail of Fig. 76.
All Allantois, *Ar* archenteron, *P* primitive streak, *V* foregut pocket. 54:1

Fig. 78 Cross section posterior to the second somite.
Beginning of somite formation, *Ch* notochordal plate.
KT 984/3, 2 somites. 270:1

Fig. 79 Cross section through the 5th somite, narrowest region of the neural groove.
Ch Notochord, *Ao* dorsal aorta, *So 5* somite 5.
KT 939/b5, 7 somites, 8 days 1 h. 560:1

Fig. 80 Low magnification of uterus, cross section, containing 2-somite embryo, cut longitudinally.
U remnant of uterine lumen.
KT 984/2, 8 days 4 h. 22:1

Fig. 81 Detail of Fig. 80 with front end of embryo.
Arrow in foregut pocket, *He* heart rudiment, *Ra* oral plate. 560:1

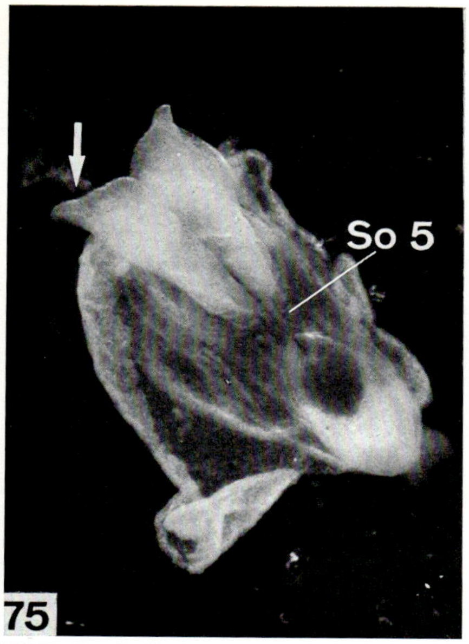
So 5
75

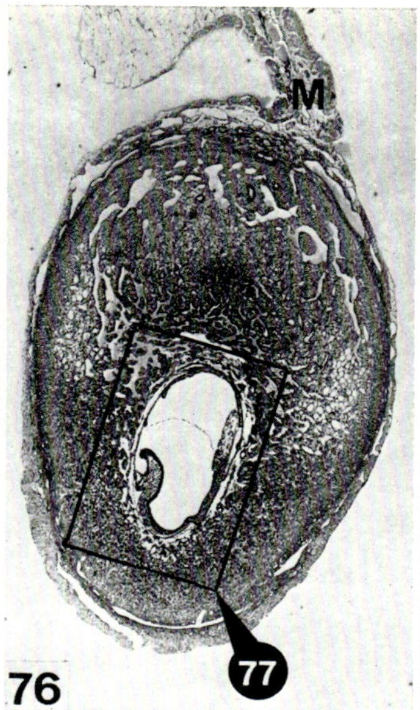
M
77
76

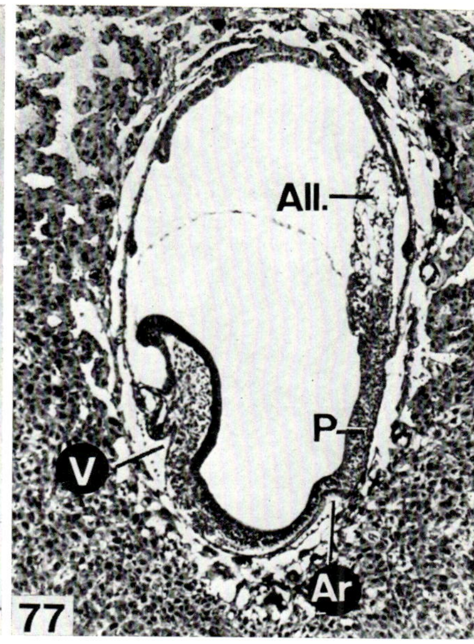
All.
P
V
Ar
77

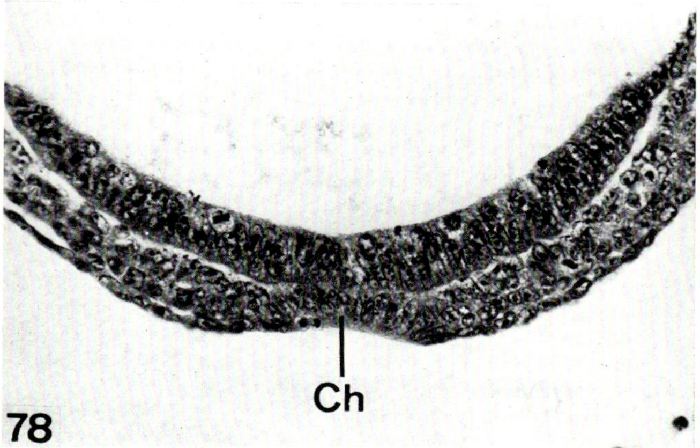
Ch
78

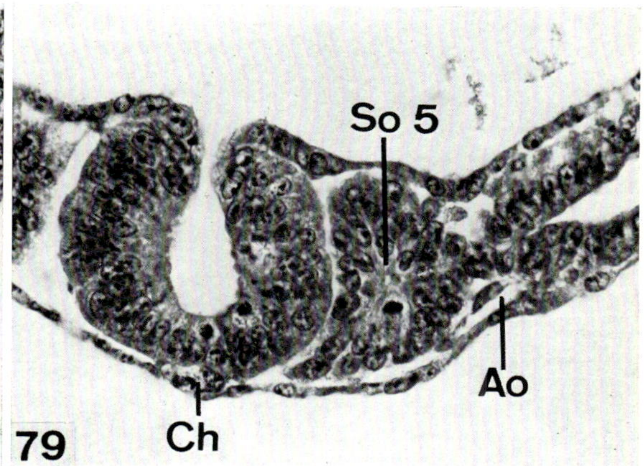
So 5
Ao
Ch
79

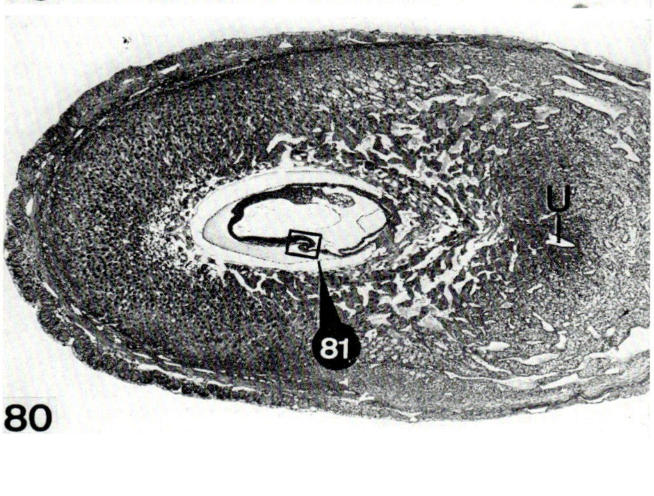
U
81
80

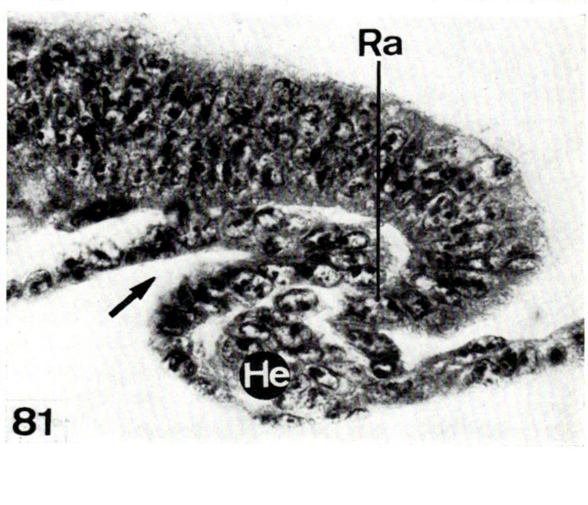
Ra
He
81

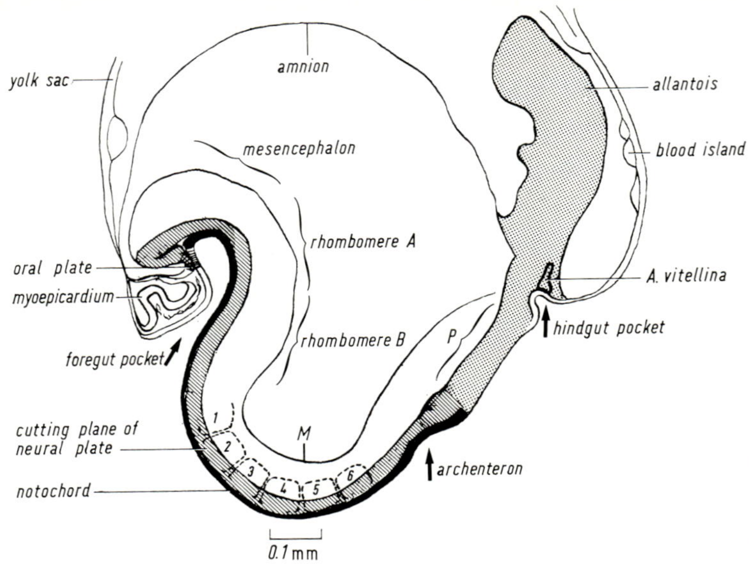

Fig. 82
Reconstruction of specimen, 8 days 1 h, 6 somites (numbered), in sagittal plane.
M Contour of right neural fold, *P* primitive streak.
KT 939/b 15

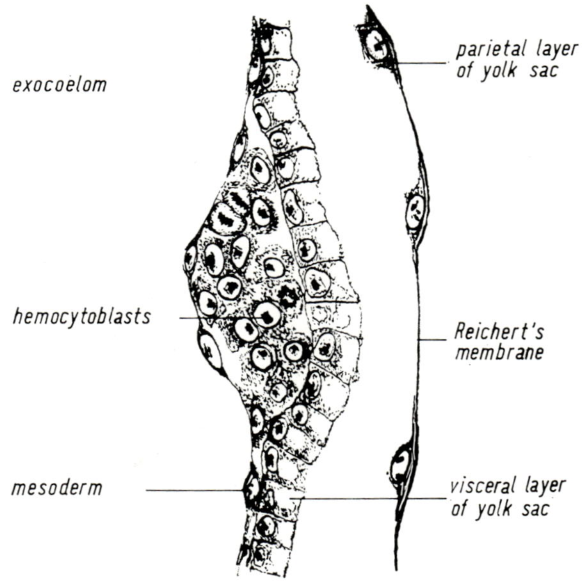

Fig. 83
Blood islet at 7 days 21 h, 5 somites.
Numerous mitoses.
KT 957/4

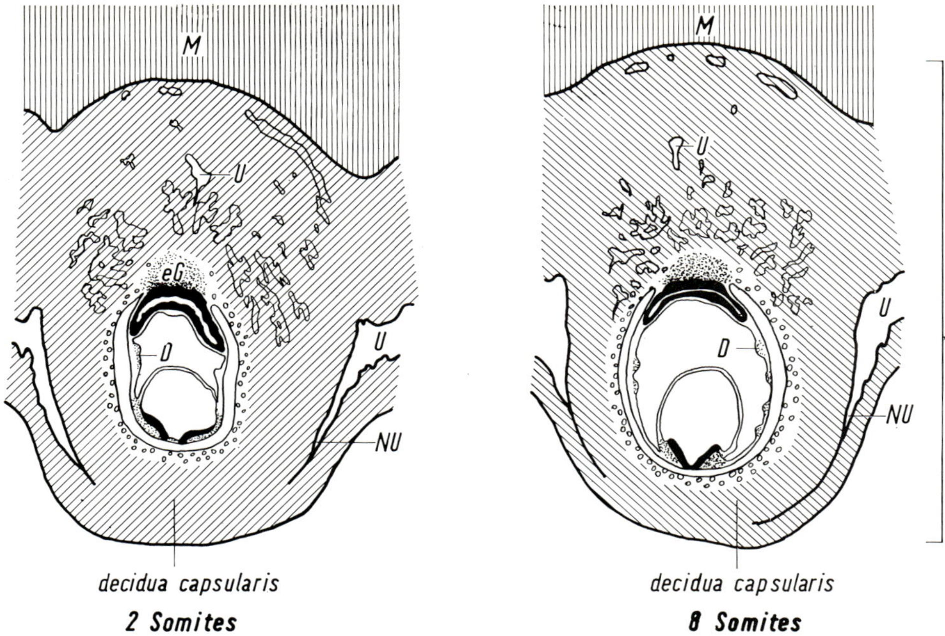

Fig. 84

Endometrium and embryonic membranes.
Maternal tissue shaded, *M* mesometrium, *U* old uterine lumen. The new lumen *NU* advances towards the antimesometrial pole, *eG* ectoplacental glycogen cells. *Circles:* trophoblastic giant cells. Enlargement of the vascularized yolk sac (*D*). Ectoplacental cavity disappears

Material	Age	
Kt 956/57	7 days 21 h	2 with 5 somites 5 with 5–8 somites 1 resorption
KT 639	8 days 1 h	6 with 5–9 somites
KT 880/81	8 days 1 h	7 early somite stages (5 somites)
KT 983/84	8 days 4 h	1 presomite neurula 2 with 1 somite 1 with 2 somites 2 with 3 somites 2 with 4 somites
KT 958/59	8 days 4 h	2 presomite neurulae 4 with 4–6 somites 3 resorptions
KT 637	8 days 21 h	3 with 4–5 somites

Horizon X/b
Older group
Early somites present

Stage 13 Turning of the Embryo

8 1/2 Days, 8–12 Somites

This is a relatively short period. The copulation age of these specimens still varies considerably. It extends from 8 days 1 hour to 9 days.

The rotation of the embryo results in a marked change of the external shape (Fig. 94). The highly lordotic curvature of the trunk turns into a strong dorsal, kyphotic bend. A primary lordotic curvature also exists in human embryos, but the kyphotic change is less conspicuous and does not involve a rotation. It is appropriate, therefore, to consider this period in the mouse development separately.

The *turning:* At 8 somites, the beginning of rotation can be seen in cross sections. It is first confined to the head and tail folds. The mid-trunk region remains initially in its original position, being apparently firmly attached to the yolk sac. In Fig. 95, the torsion of the posterior end with its primitive streak with respect to the mid-trunk region is apparent. Viewed from the cranial towards the caudal end, the rotation proceeds clockwise along the body axis.

Figs. 85–93: 8 somites, 8 days 21 h

Fig. 85 Reconstruction of embryo, 2 somites, 8 days 21 h.
Level of cross sections Fig. 88–92 is indicated.
KT 985

Fig. 86 Ovary with 2 corpora lutea of the same pregnancy.
Cl Corpus luteum.
KT 985. 16:1

Fig. 87 Cellular detail. 130:1

Fig. 88 Anlage of the forebrain.
Arrow: sulcus opticus. 180:1

Fig. 89 Anlage of the heart.
End Endocardiac vesicle, *Tr* aortic sac, *Th* Plate of the thyreoidea. 130:1

Fig. 90 Foregut pocket (*V*).
S.v. Sinus venosus (paired), *O* otic plate (posterior margin). 130:1

Fig. 91 Section through 1st somite (*So.1*).
V.v. Vitelline vein. 130:1

Fig. 92 Section through 4th somite (*So.4*).
Ao Aorta dorsalis, *Pt* peritoneal funnel. 270:1

Fig. 93 Section through primitive streak (*P*).
H Hind gut, *Au* umbilical artery. 270:1

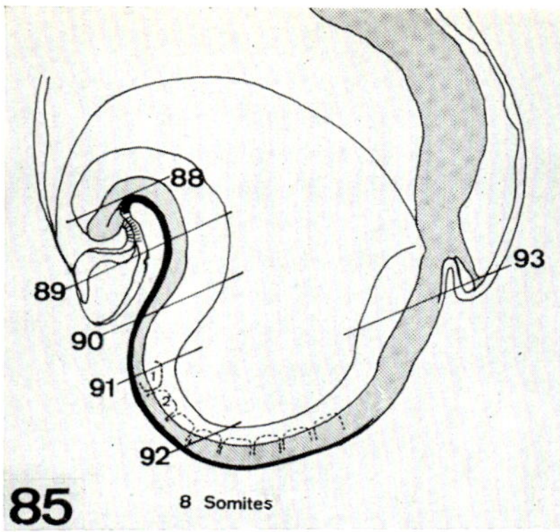
88
89
90
91
92
93
8 Somites
85

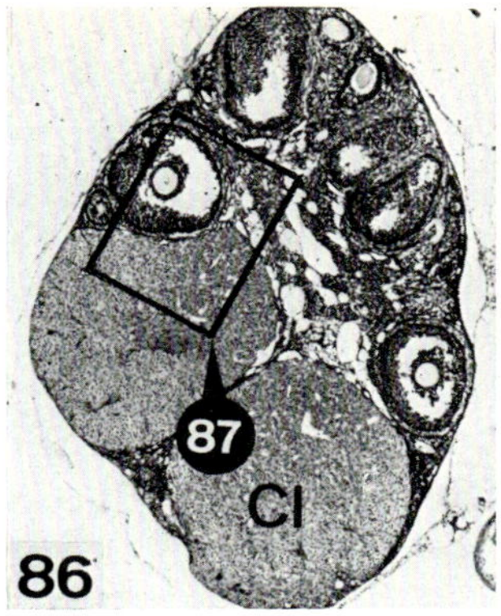
87
Cl
86

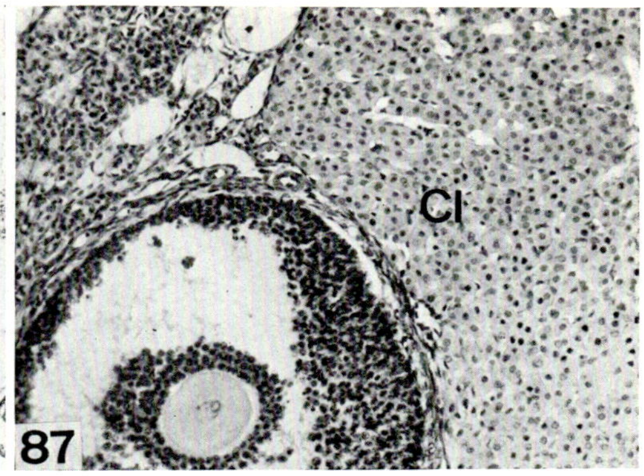
Cl
87

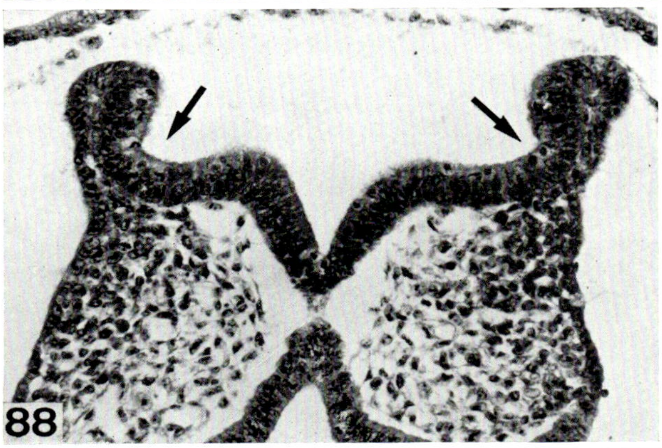
88

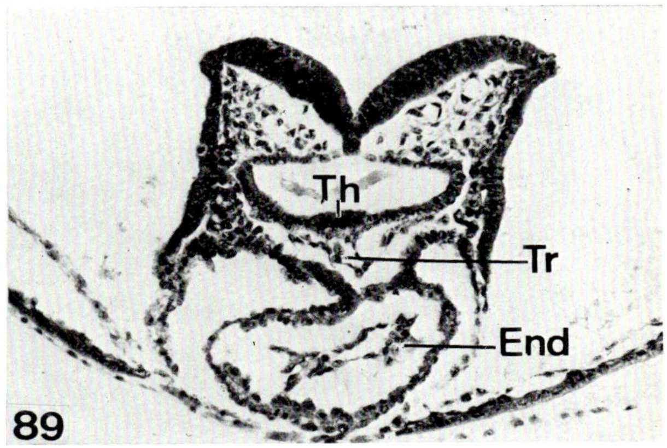
Th
Tr
End
89

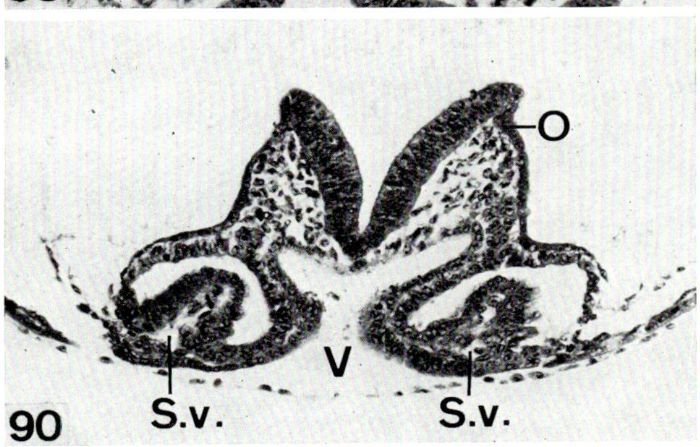
O
V
S.v.
S.v.
90

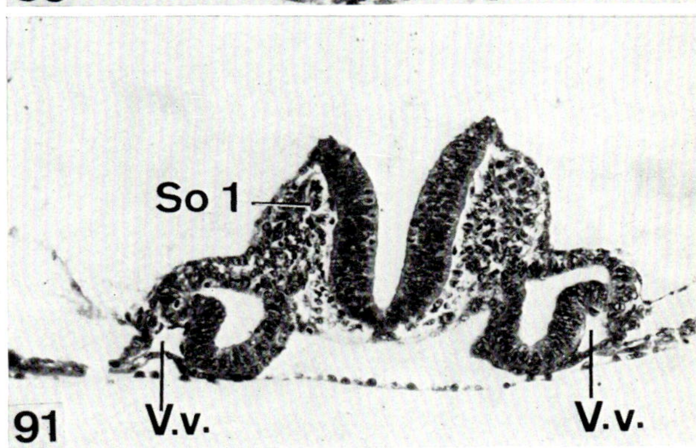
So 1
V.v.
V.v.
91

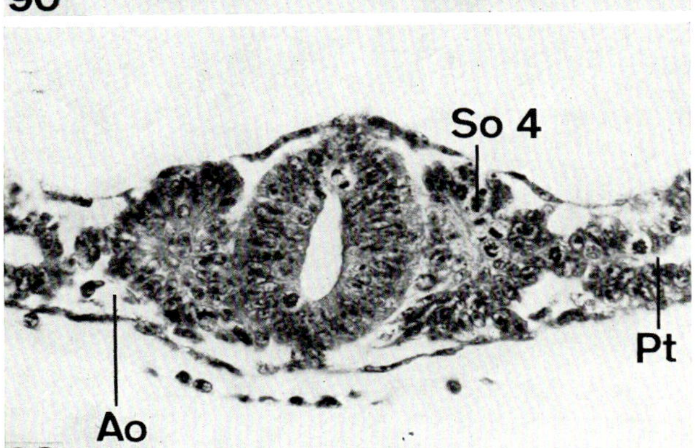
So 4
Pt
Ao
92

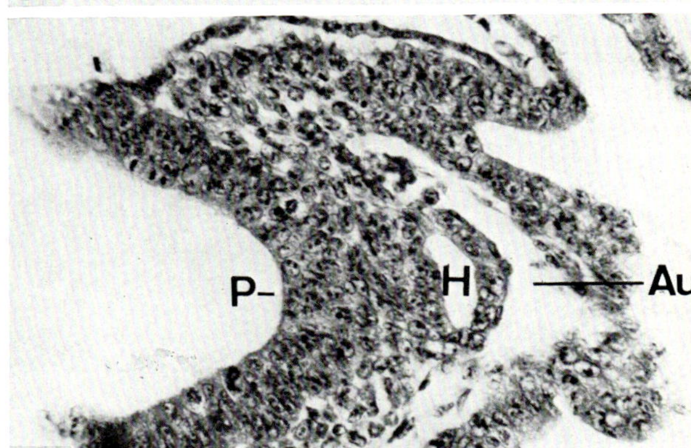
P
H
Au
93

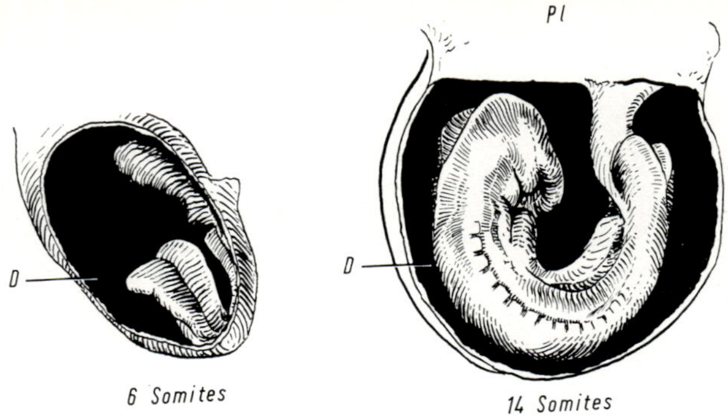

Fig. 94
Turning of the embryo.
Drawing before (left) and after (right) rotation. Amnion is not shown. *D* Yolk sac (cut), *Pl* placenta

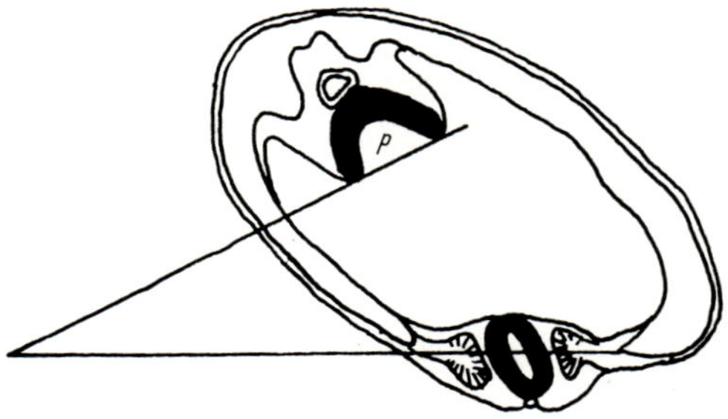

Fig. 95
Beginning rotation, 8 somites.
P Tangential line connecting the primitive folds, shows the turning compared to the trunk

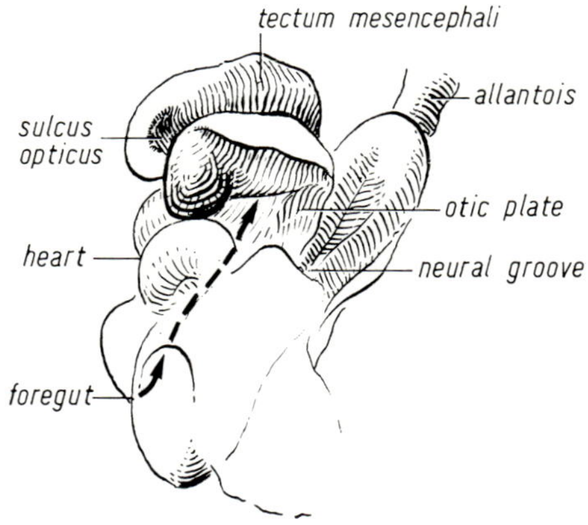

Fig. 96
Folding of the brain bulges. Oblique view.
Arrow: localization of the foregut.
KT 1002, 9 somites, 8 days 10 h

Fig. 97
Dorsal view of the brain folds. They have approached each other closer in the fore-than in the hindbrain.
KT 1002, 9 somites

Organogenesis

Compared to the preceding period, no radical changes occur during this stage. The previously described organ rudiments may easily be recognized, and are represented in a series of cross-sections of an embryo of 8 somites (Figs. 85–93). The optic evagination is deeper now and the otic plate more distinct (Fig. 96). The thyroid rudiment [144] is clearly delimited (Fig. 89). In sagittal sections, it appears as an indentation of the fore-gut wall above the heart rudiment (Fig. 85).

At the end of this period, the second branchial pouch is forming. At 8 somites the first pouch is considerably enlarged, so that the entoderm contacts the overlying ectoderm. The section in Fig. 89 is immediately anterior to the contact area. The fore- and hindgut-pockets are deeper now, and in the region of the somite stalks, the peritoneal funnels of the pronephros are sometimes visible (Fig. 92).

The Corpus Luteum

The structure of the corpus luteum is practically unaltered; there is little change, even when compared with the five-day-stage (Fig. 36). The central scar and hemorrhagic traces have almost disappeared. Sometimes, a zone of loosely arranged cells is observed in this area. The histologic picture of the corpus luteum and of the interstitial gland is still unchanged.

Material	Age	Embryos
KT 639	8 days 1 h	6 with 5–9 somites (mentioned previously)
KT 1001–03	8 days 10 h	2 with 7 somites 2 with 8 somites 1 with 9 somites 2 resorptions
KT 985–87	8 days 21 h	7 with 8–16 somites

Stage 14 Formation and Closure of Anterior Neuropore

9 Days, 13–20 Somites

Horizon XI
13–20 paired somites

This period could be extended to 22 somites, because the most advanced embryo, with a copulation age of 9 days 2 hours, had 23–24 pairs of somites. In the mouse, the formation of a new somite pair requires 1–2 hours. I have chosen an upper limit of 20 somites for this stage so that it will correspond to Streeter's horizon XI.

External Form

The turning of the embryo is now complete. Compared to human embryos, the mouse embryos of this group are strongly flexed in a dorsally convex direction (Figs. 99, 108). Furthermore, there is a definite spiral torsion, the posterior end lying on the right side of the head. In rare cases, it may be found on the left side.

The external form is determined in this phase basically by the shape of the neural tube. Anteriorly, the medullary plate is about to close, whereas it is still a flat groove posteriorly. The first two branchial bars are clearly visible. The forelimb bud is not yet clearly delimited. At 15 somites, it appears as a condensation of the lateral plate material, without distinct boundaries (Fig. 100).

Length: The overall length varies considerably, mainly because of varying curvature of the body. If the amnion is cut, the embryo straightens a little. The length is measured from the crown to the curved posterior end in a straight line, and it varies from 1.2–2.5 mm. Externally, the roundish swellings of the uterus measure 3–5 mm. The spaces between the embryos are irregular in vivo and if the mouse is killed they are reduced by contraction of the uterus [26].

Circulatory System

In transparent fresh or formalin-fixed embryos, several vessels may be recognized by simple inspection (Fig. 108).

The broad *anterior cardinal vein* receives its blood from the close-meshed network of capillaries investing the neural tube (plexus perineuralis). It joins the posterior cardinal vein to form the *Ductus Cuvieri.*

The (paired) dorsal aorta may be seen immediately ventral to the row of somites. The endocardial tube of the bulbus cordis is very narrow and is separated by a considerable gap from the thick myoepicardium. Anteriorly it forms a right angle with the arterial trunk which is dilated at its end (aortic sac).

Development of the heart. The heart is now capable of maintaining some circulation of the blood. The atrium and ventricle are not yet paired. The shape of the arterial and venous parts of the heart, together with the connecting vessels, are shown in Figs. 109 and 110.

Placental circulation is just being established. The blood already circulates in the yolk sac. The paired dorsal aortae supply the yolk sac by means of a thick vitelline artery (Fig. 111). This artery will originate later separately from the aorta by a new anastomosis [49], whereas the umbilical artery will remain in direct continuation of the aorta. This transformation is illustrated in Fig. 112.

Intestinal Tract

The original wide opening of the gut into the yolk sac is narrowed to a long, slender *vitelline duct*. As a consequence, fore- and hindgut are no longer represented by separate pockets, but form a continuous tube with blind ends. In the middle, it still opens into the yolk sac.

At 16 somites, the *oral plate* may rupture. At 21 somites only remnants of the membrane remain. The *cloacal membrane*, on the other hand, does not appear as a distinct membrane until the 16-somite-stage; and it persists for a long time.

The *foregut* is now differentiating rapidly. The first two branchial clefts have formed (Fig. 113). The floor of the foregut is thickened, while the dorsal epithelium remains thin. Ventrally, the *thyroid rudiment* evaginates, forming a small groove (Fig. 104). It lies just above the dilated aortic sac. Toward the end of this period, the *lung anlage* (rudiments of larynx–trachea–bronchi) appears as a ventral thickening of the epithelium. There is only a short distance between the lung rudiment and the hepatic diverticulum. This liver primordium develops earlier in mice than in humans and rapidly deepens. The stomach has not yet formed.

The coelom is a single cavity, the cavum pleuro-pericardiaco-peritoneale. In the vicinity of the umbilical ring it communicates with the exocoelom. The coelomic epithelium is thickened above the lung rudiments and in the region of the future stomach [83].

Urogenital System

In this period the somite stalks of the lower cervical and upper thoracic regions are composed of:

1. Peritoneal funnel,
2. Nephric vesicle; sometimes its lumen is still lacking,
3. Nephric duct, as a rule still solid (Fig. 102).

At 15 somites, this typical organization can be observed at the level of the 10th to the 14th somite (KT 987). In another specimen, KT 986, with 16 somites, the pronephric duct has developed a distinct lumen at the level of the 15th somite. The anterior somites, however, are connected to the coelomic epithelium only by disorganized clusters of cells.

Germ cells [103] can be recognized in H.-E.-sections for the first time at 15 somites, within the epithelium of the hind gut (KT 987).

Central Nervous System

The formation and closure of the anterior *neuropore* is a most remarkable event. It is accomplished at the same developmental phase as in humans, but seems to be displaced caudally. A comparison of Fig. 75 (7 somites) and 96 (9 somites) reveals an irregular closure of the brain folds. Initially, at 7 somites, they approach each other only posteriorly. Later, they also approach each other anteriorly in the forebrain region, while there is still a wide gap in the

Figs. 98–107: Closure of anterior neuropore, 9 days, 15 and 21 somites

Fig. 98 Embryo in yolk sac (*D*) and amnion (*A*).
KT 988, 15 somites, 8 days 21 h. 7:1

Fig. 99 Dissection of embryo from membranes.
O Otic invagination.
KT 988, 15 somites. 11.5:1

Fig. 100 Oblique view of same embryo, showing anterior neuropore (*V.N.*).
Aa Forelimb. 10.5:1

Fig. 101 Cross section through embryo, 15 somites, 8 days 21 h.
Ao Aorta, *Au* umbilical artery, *H* hindgut.
KT 987/1. 180:1

Fig. 102 Detail of Fig. 101, with somite 14.
Ao Aorta; *Vn* pronephric duct, without lumen here; *Pt* peritoneal funnel. 350:1

Fig. 103 Cross section through same embryo, 15 somites, at the level of the facial nerve (*N. VII*).
V.c. Vena capitis lateralis, *Ao* aorta, *V* foregut.
KT 987/1. 130:1

Fig. 104 Detail of thyreoidea-anlage (*Th*).
Tr Truncus arteriosus (aortic sac)

Fig. 105 Detail of Fig. 106, showing boundary zone of ectoplacental cone.
D Yolk sac, *R* Reichert's membrane, *T* trophoblastic giant cells, *All* allantois. Zone of contact with embryonic vessels, *eG* ectoplacental glycogen cells, *De* decidua basalis. 100:1

Fig. 106 Cross section through uterus, with embryo and extraembryonic membranes.
A Amnion, *Au* umbilical artery in umbilical cord.
KT 624, 9 days, 21 somites. 18:1

Fig. 107 Detail of Fig. 106
showing decidua capsularis and new uterine lumen (below).
T Trophoblastic giant cells, forming meshwork; *R* Reichert's membrane and distal (parietal) layer of yolk sac; *D* yolk sac, proximal (visceral) layer, with blood-vessels. 100:1

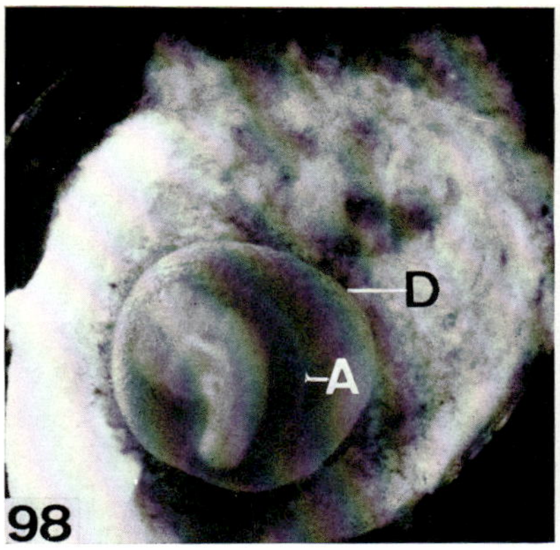

98

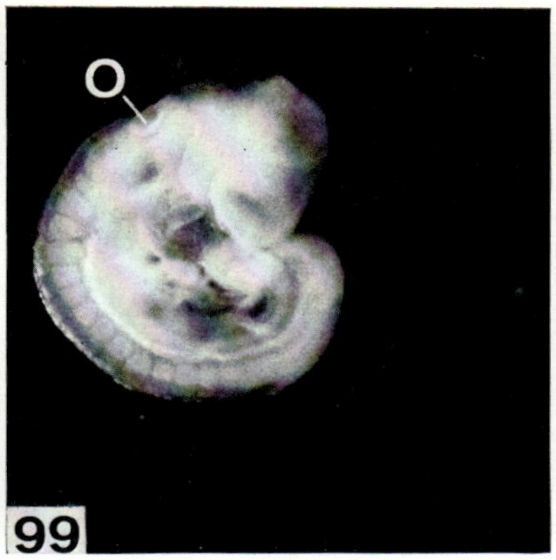

99

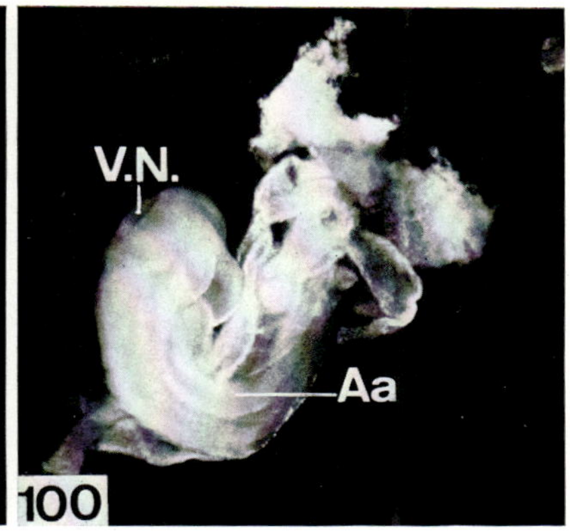

100

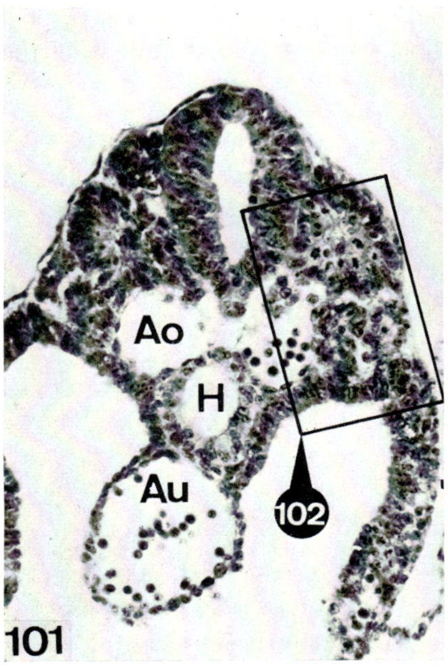

101

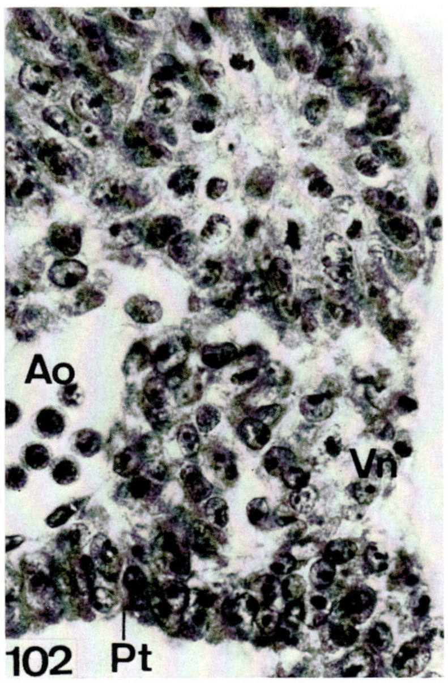

102

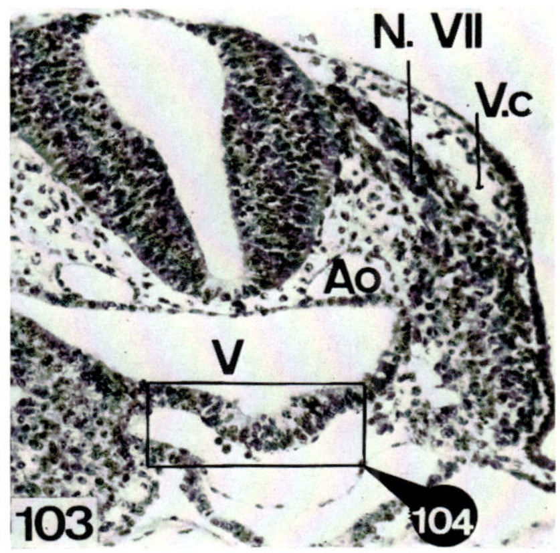

103

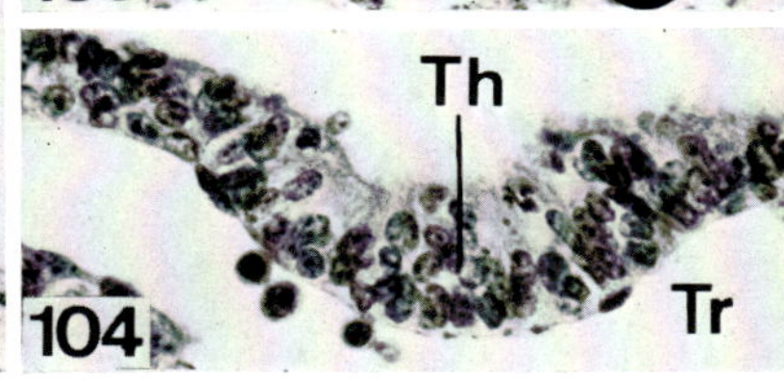

104

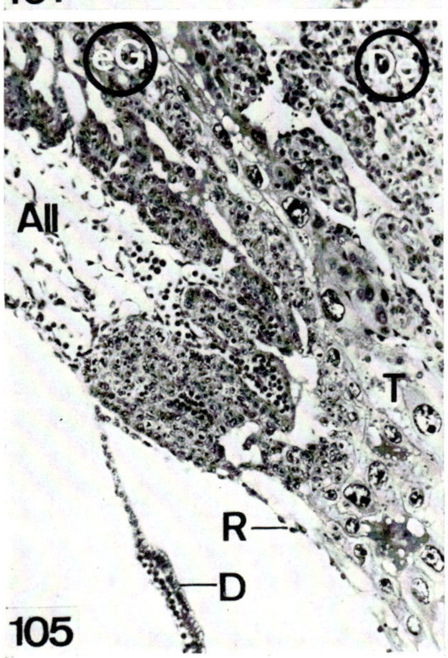

105

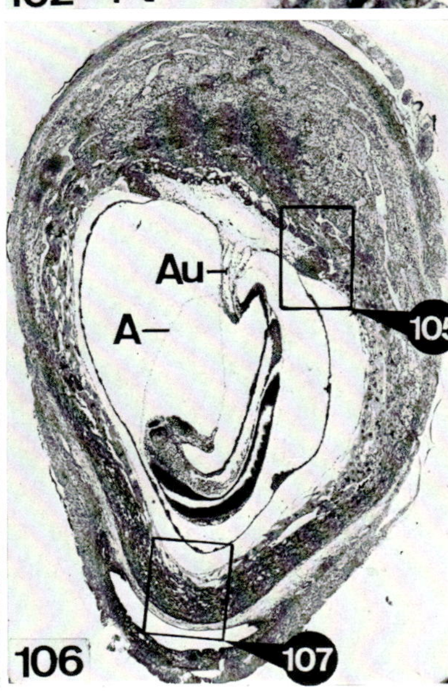

106

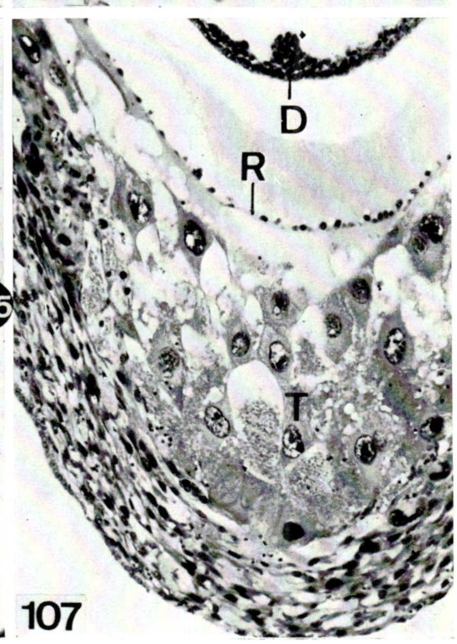

107

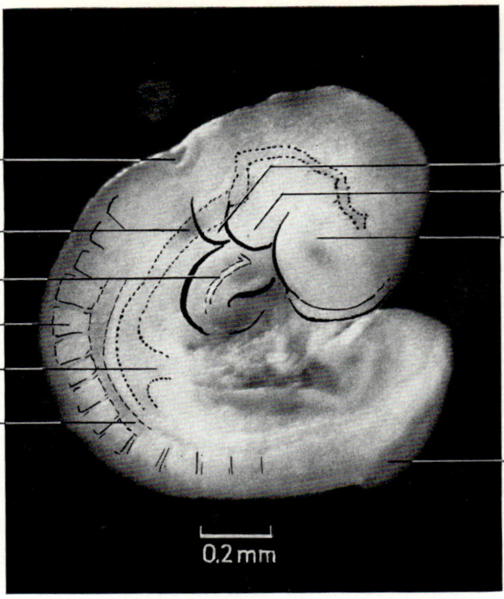

Fig. 108
Embryo, formalin fixed.
KT 937a, 14 somites, 9 days

anterior rhombencephalon (Figs. 97 and 100). They do not close in this region until 15–18 somites have formed. The closure does not coincide with the attainment of a definite somite number. The earliest was observed at 15 somites. In any case, the folds are closed at 19 somites. During dissection, a newly closed neural tube may open again, even if the uterus is fixed in toto.

After the closure of the anterior neuropore, the floor of the rhombencephalon acquires a characteristic shape. Six *rhombomeres* [162] are formed which will disappear later. Prior to the closure of the neural tube, the rhombomeres may be seen in different phases of maturation (compare Fig. 82).

The *cranial neural crest* [161] is visible *prior* to the closure of the brainfolds. The trigeminal- and facialis crests are most distinct. In Fig. 103, the rudiment of the facialis-ganglion may be recognized. The overlying epidermis is thickened and represents the facialis-placode.

The *eye anlage* is in the vesicular stage. The optic evagination reaches the overlying epidermis prior to the closure of the neural tube and induces the formation of the lens placode. This placode can be easily recognized by the tall epithelial cells which are formed shortly after the disappearance of the anterior neuropore.

The *olfactory placode* appears shortly before the lens placode, as a striking thickening of the epithelium, while the brain tube is still open. At first it is adjacent to the forebrain. Now it begins to be separated from it by invading mesenchyme.

The *otic plate* is transforming into a deep open groove. The invagination proceeds more or less regularly with increasing somite numbers.

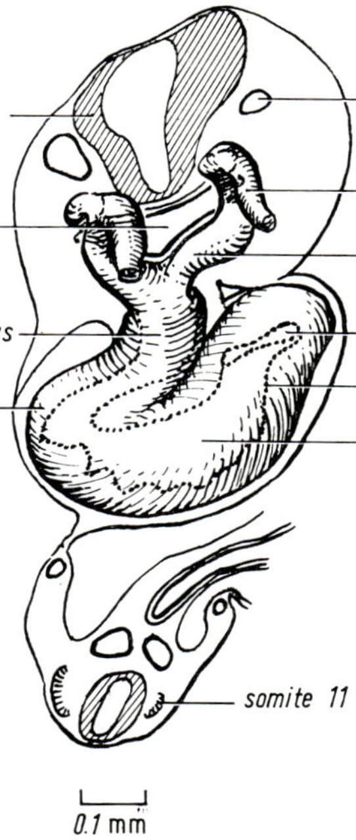

Fig. 109
Reconstruction of the arterial region of the heart.
Ventral view, based on a frontal section through somite 11 and anterior rhombencephalon.
KT 987, 16 somites, 8 days 21 h

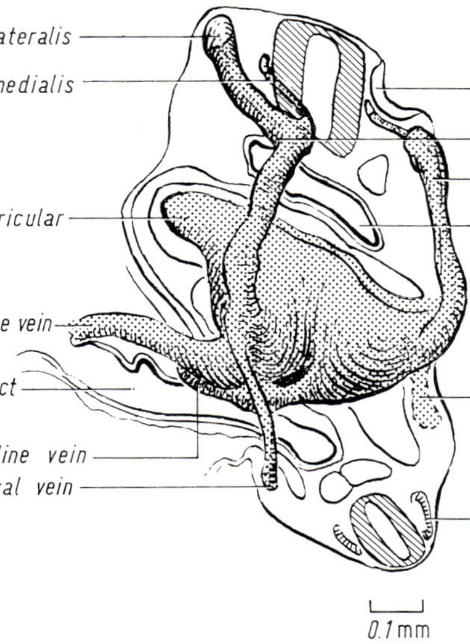

Fig. 110
Reconstruction of the venous sinus.
Dorsal view, from left side. Based on a frontal section, level of somite 9 and otic invagination. The venous sinus lies dorsal to this plane.
KT 987, 16 somites, 8 days 21 h

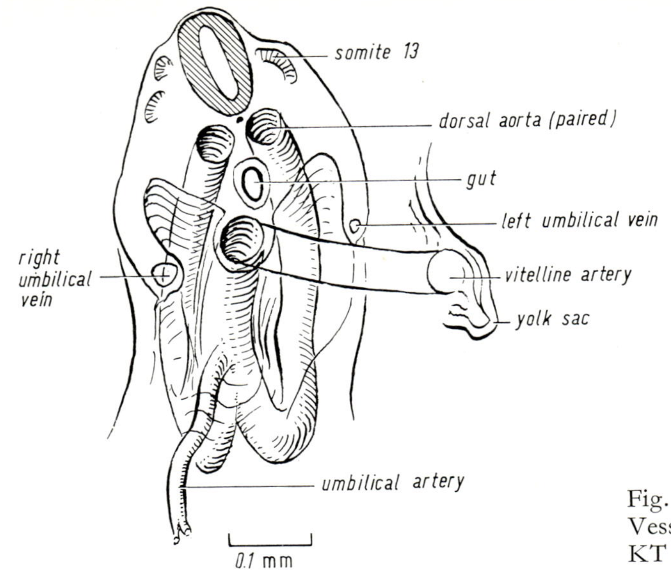

Fig. 111
Vessels of the posterior body region.
KT 987, 16 somites, 8 days 21 h

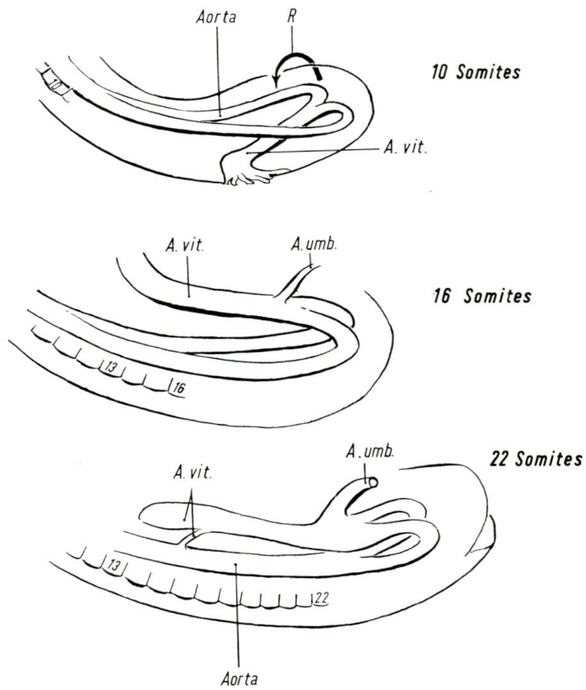

Fig. 112
Development of the posterior arterial stems. At 22 somites, the vitelline artery takes a new origin, by means of an anastomosis. The connection with the umbilical artery is lost later. The *arrow* (*R*) indicates the direction of the embryonic rotation

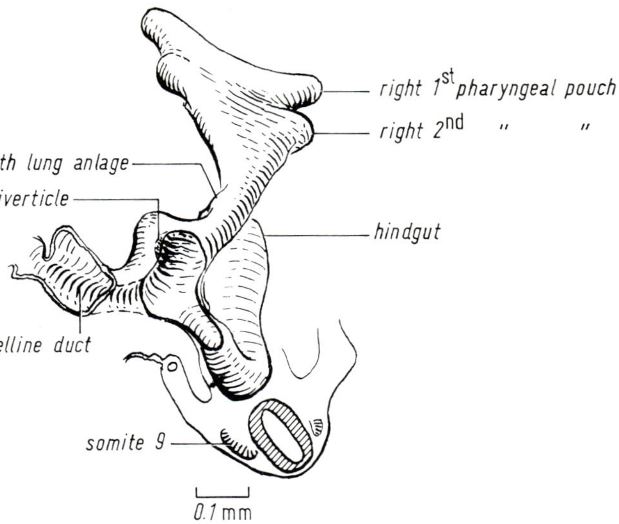

Fig. 113
Anlage of the gut, viewed from dorsal and left, starting from a section at the level of the 9th somite.
8 days 21 h, 16 somites

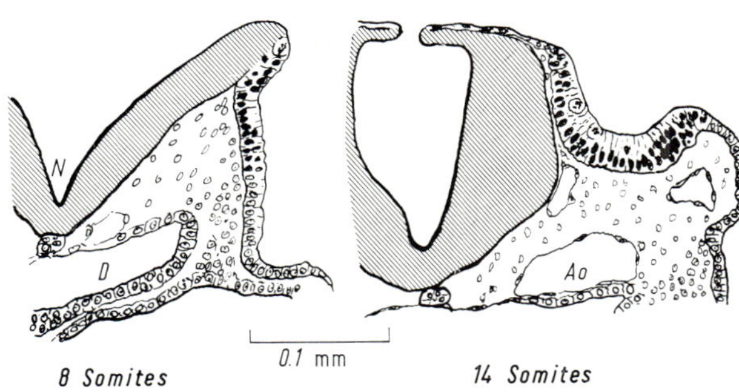

Fig. 114
Development of the otic invagination. Cross sections. The neural tube remains open here for a longer time than in human embryos. *D* Gut-anlage, *Ao* aorta, *N* neural groove

Endometrium and Placenta

The new uterine lumen (Fig. 84) is now continuous (Fig. 106). The adjacent decidua capsularis is poorly vascularized and is composed of relatively small cells. It borders a zone of trophoblastic cells which form a perivitelline meshwork invaded by maternal blood (Fig. 107). Near this zone, the deciduous cells are often multinucleate and are darkly stained.

The *placenta* is developing in the region where the allantois reaches the ectoplacental plate (= "chorionic plate"). The ectoplacental cavity disappears by fusion of the two ectoplacental laminae. The fusion begins in the middle and reaches the margins by 8 days (Fig. 77). Towards the maternal tissue, the boundary zone is organized into two different regions:

1. Toward the embryo, there is a rather solid wall of cells, into which the large allantoic vessels enter. They contain some nucleated embryonic erythrocytes (Fig. 105).

Toward the decidua numerous clefts are developing so that the cell mass is split into anastomosing strands. The clefts are filled with maternal blood. In this way, the *labyrinth* of the placenta is formed.

2. Further toward the decidua basalis there is a zone of ectoplacental glycogen cells which are joined, more peripherally, by single trophoblastic giant cells, in continuation of the peri-

vitelline meshwork (Fig. 105). In contrast to the placental labyrinth, only *maternal blood* is circulating in this junctional zone [38] (= Trophospongium = reticular zone).
The yolk sac is provided now with a well formed vascular net, and it functions as additional "placenta" (yolk-sac placenta). In the course of dissection, the visceral layer of the yolk sac is immediately exposed after cutting the uterine muscular wall and decidua. There is a cleft between the proximal and distal layer, and the distal layer adheres to the decidua.

Material	Age	Embryos
KT 624–26	9 days	1 with 5 somites 1 with 11 somites 1 with 16 somites 2 with 20 somites 1 with 21 somites 1 resorption
KT 985–88	8 days 21 h	1 with 8 somites, anterior neuropore open 1 with 13 somites, anterior neuropore open 2 with 14 somites, anterior neuropore open 1 with 15 somites, anterior neuropore open 2 with 16 somites, anterior neuropore open 1 resorption
KT 935–37	9 days 3 h	1 with 15 somites 2 with 17 somites, anterior neuropore closed 1 with 20 somites, anterior neuropore closed 3 with 22 somites, anterior neuropore closed 1 with 23 somites, anterior neuropore closed 1 with 24 somites, anterior neuropore closed

Stage 15 Formation of Posterior Neuropore; Forelimb Bud

9 1/2 Days, 21–29 Somites, 1.8–3.3 mm

Horizon XII
21–29 somites,
homo = 3–4 mm

External Shape

During this phase, the condensation of the *forelimb bud* may become apparent for the first time. It is situated at the level of the 8th–12th somite (Fig. 116). A distinct condensation of the hind limb bud does not appear until the end of this period.
A prominent feature is the presence of *3 branchial bars* compared to two in the preceding age group.
The *optic vesicle* is still spherical and has not yet begun to invaginate. In Figs. 115–116, the central light spot represents the wide open stalk of the optic vesicle and not the lens rudiment.
The *otic vesicle* is usually closed, in any case from the 24-somite stage onward.
The anterior *neuropore* is invariably closed, while the posterior one is open in all specimens examined. Compared to humans the closing of the posterior neuropore is thus retarded in mice and does not start before the next period.
Length: The extent of curvature of the embryo varies considerably, and therefore the length varies. In the fresh state, they range from 1.8 to 3.3 mm. After fixation in formalin, followed by storage in 70% alcohol, length is reduced by nearly one-third.

Sagittal Section (Fig. 121)

Fig. 121 shows the relationships between some of the internal organs. Rathke's pouch is still open. The liver anlage is still covered by the heart.

Circulatory System

The *heart* consists of a convoluted tube, not yet divided into right and left parts. Through its transparent myocardial wall, the fine endocardial tube may be recognized (Fig. 118). The cardiac gelatinous coat (= subendocardial jelly) contains only a few cells and is in the process of forming the atrio-ventricular cushions (Fig. 122).
The first *aortic arch* is of small caliber. The second and third are well developed (Fig. 122). They conduct the blood into the paired dorsal aortae and through the umbilical artery into the placenta. The vitelline artery (= A. omphalo-mesenterica) develops a new origin from the dorsal aorta by way of a newly developed anastomosis [49] (Fig. 112).
The sinus venosus receives the same tributaries as in the previous stage (Fig. 110). However, it gradually separates from the atrium proper (atrium commune) by the development of a transverse ridge (Fig. 122).

Intestinal Tract

The intestinal tract is now undergoing important transformations which are shown in Fig. 123. The *lung rudiments* appear at the beginning of this period as epithelial thickenings which are now delimited posteriorly. The laryngo-tracheal groove deepens and begins to detach from the intestinal canal.

The *stomach primordium* is expanding rapidly. At the same time the omental bursa develops as a lateral peritoneal pocket (Fig. 124).

Within the *hepato-duodenal field*, columns of cells (hepatic epithelial cords) continue to invade the mesenchymal tissue of the septum transversum.

The *pancreas* develops from two separate areas of the duodenal epithelium. The anlage of the ventral pancreas is a small, circumscribed ventral evagination in the caudal part of the hepato-duodenal field. The dorsal pancreas, on the other hand, is a broad evagination in the dorsal half of the duodenal epithelium, which is not constricted until the end of this period (Fig. 123). It is not situated cranially to the ventral rudiment as in man. The reconstruction of the 10 1/2 day stage (Fig. 123) does not show the entire intestinal tract because of a lateral curvature of the embryo.

The *vitelline duct* is now closed, and its epithelium has disappeared.

Urogenital Tract

The nephric duct has grown far into the mesonephric region (Fig. 122). As an example, in specimen KT 997/5 (25 somites), it extends from the 11th to the 25th somite and parts of it have a distinct lumen. In specimen KT 939 (28 somites), the nephric duct has reached the vicinity of the cloaca (Fig. 126).

Figs. 115–120: Foreleg-bud, 9 1/2 days, 22 and 24 somites, 1.8 and 2.9 mm length

Fig. 115 — Embryo of 22 somites, formalin fixed, translucent.
Rh Rhombencephalon with 4th ventricle, *O* ear vesicle, separating from epidermis. 36:1

Fig. 116 — Same embryo, surface illumination.
Ab Eye vesicle, *Aa* forelimb-bud, *1* and *2* first and second branchial bars. 36:1

Fig. 117 — Same embryo, from left, translucent.
Hn Posterior neuropore. 30:1

Fig. 118 — Embryo of 24 somites, formalin fixed, on millimeterscale.
O Ear vesicle, separating from epidermis; *Aa* forelimb-bud; *N* suture line of neural folds; *E* endocardium within bulbus arteriosus; *1*, *2*, *3* branchial bars 1–3. 21:1

Fig. 119 — Cross section of uterus with embryo of 22 somites.
Aa Forelimb-bud, *Ri* olfactory placode, *D* yolk sac, visceral layer. 16:1

Fig. 120 — Detail of Fig. 119.
Eye vesicle with emigrating cells of the ophthalmic neural crest (*Op*). 200:1

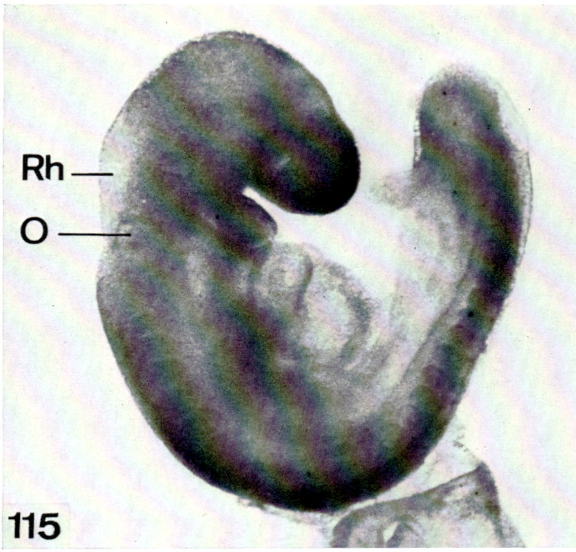
Rh
O
115

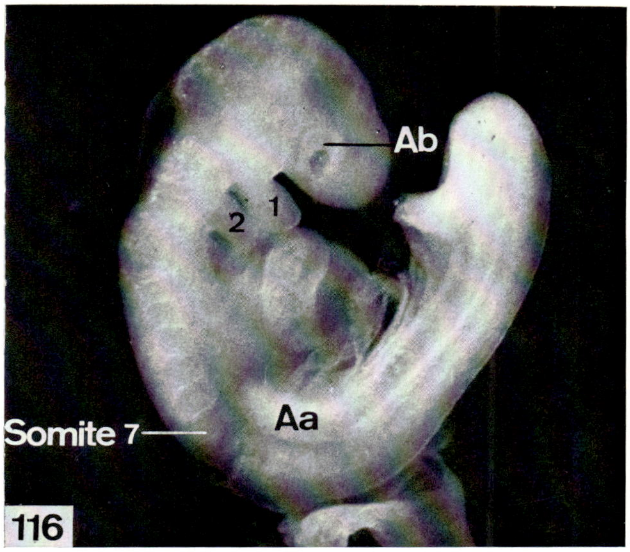
Ab
1
2
Somite 7
Aa
116

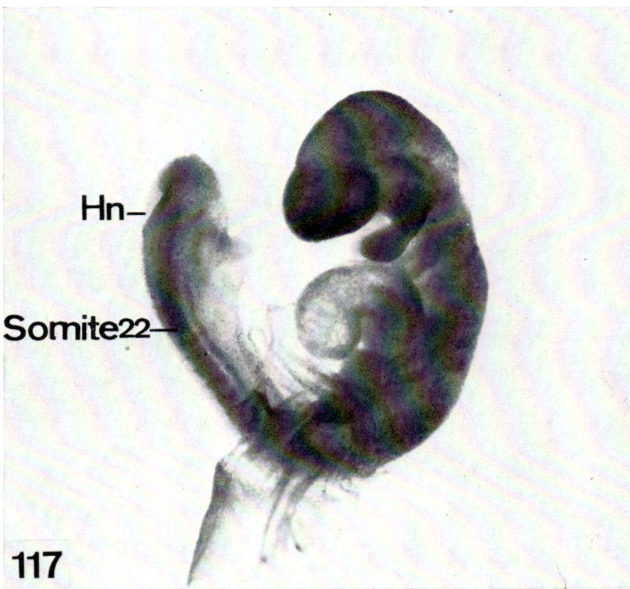
Hn
Somite22
117

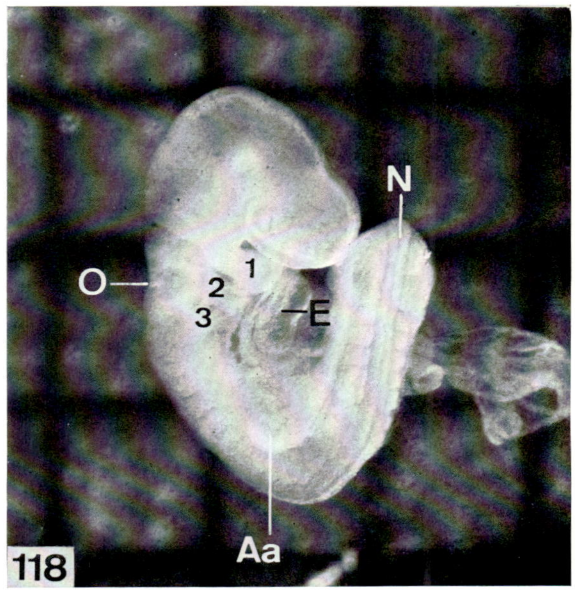
N
O
1
2
3
E
Aa
118

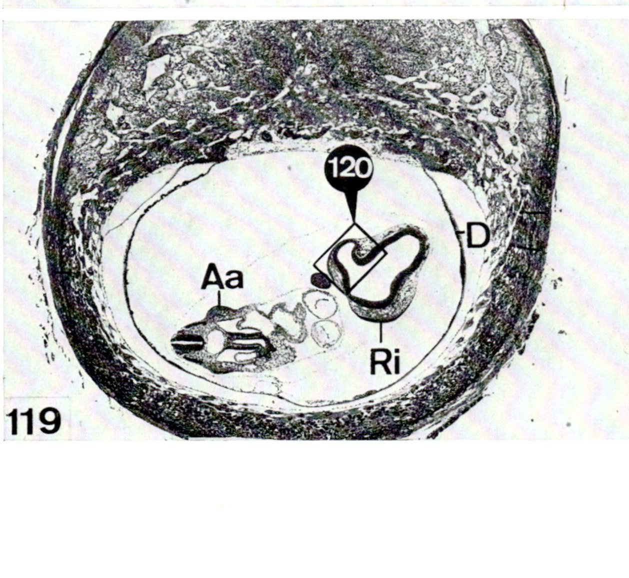
120
D
Aa
Ri
119

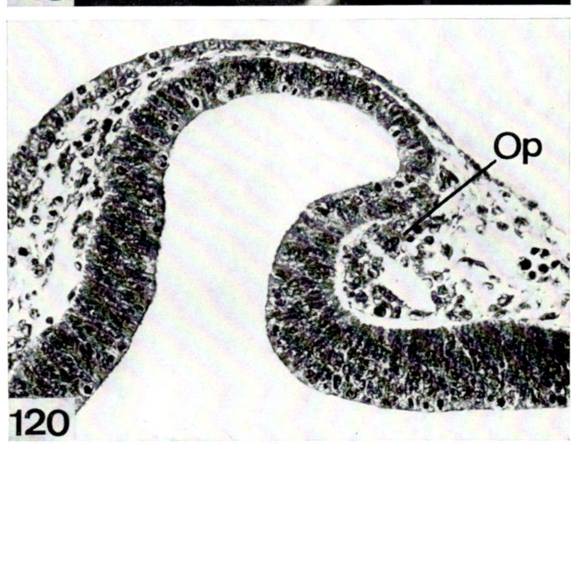
Op
120

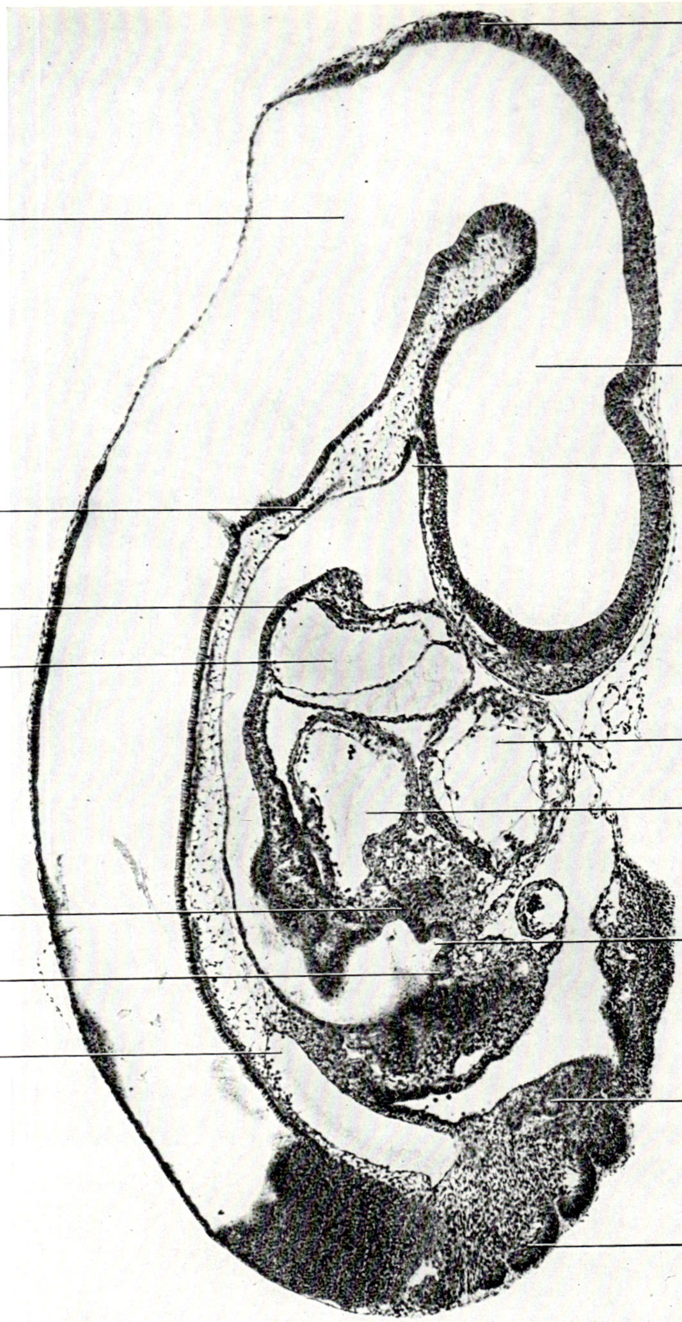

Fig. 121
Sagittal section through embryo of 10 days, 26 somites.
KT 939. 62:1

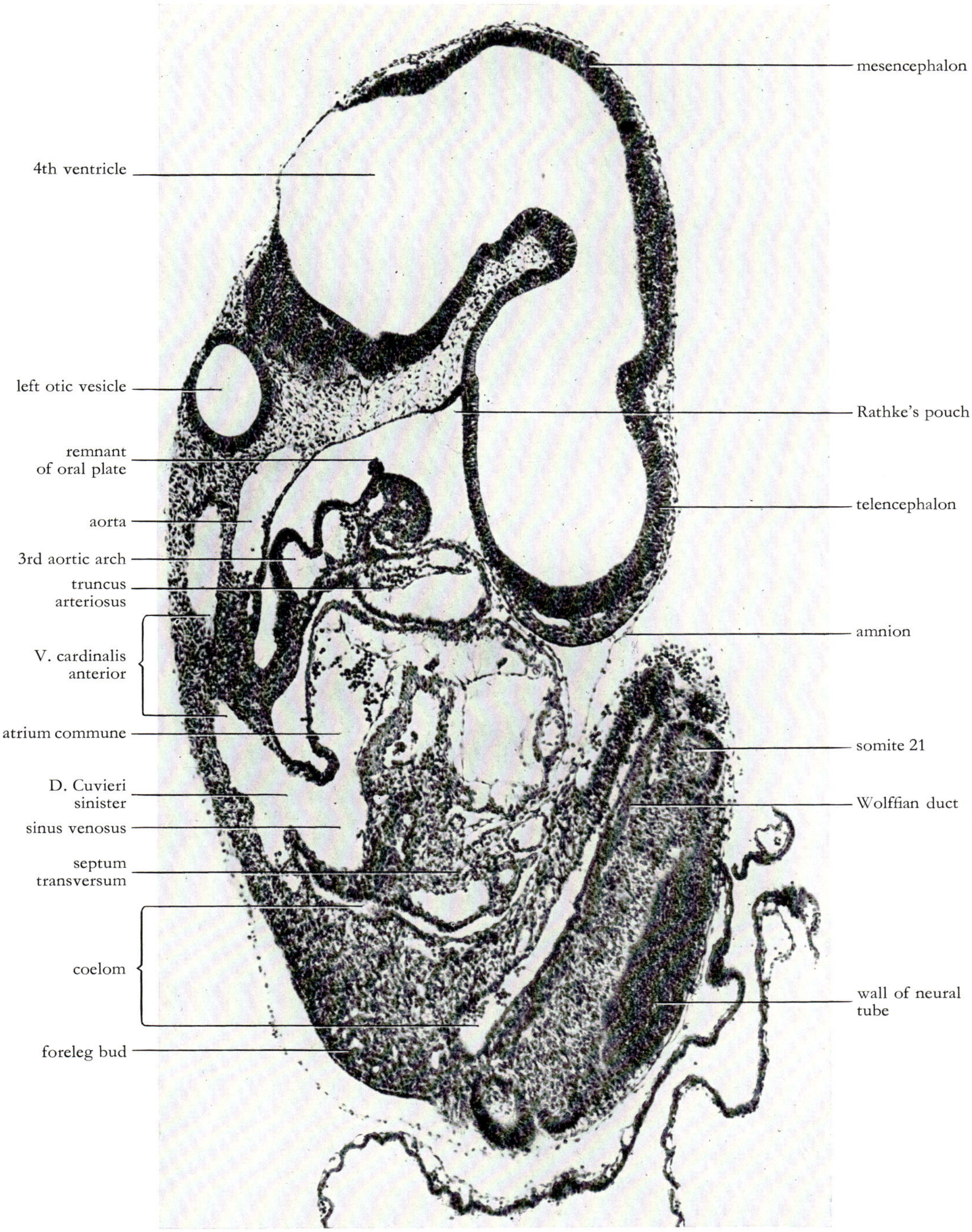

Fig. 122
Development of the circulatory system. Parasagittal section through mouse embryo of 9 days 3 h

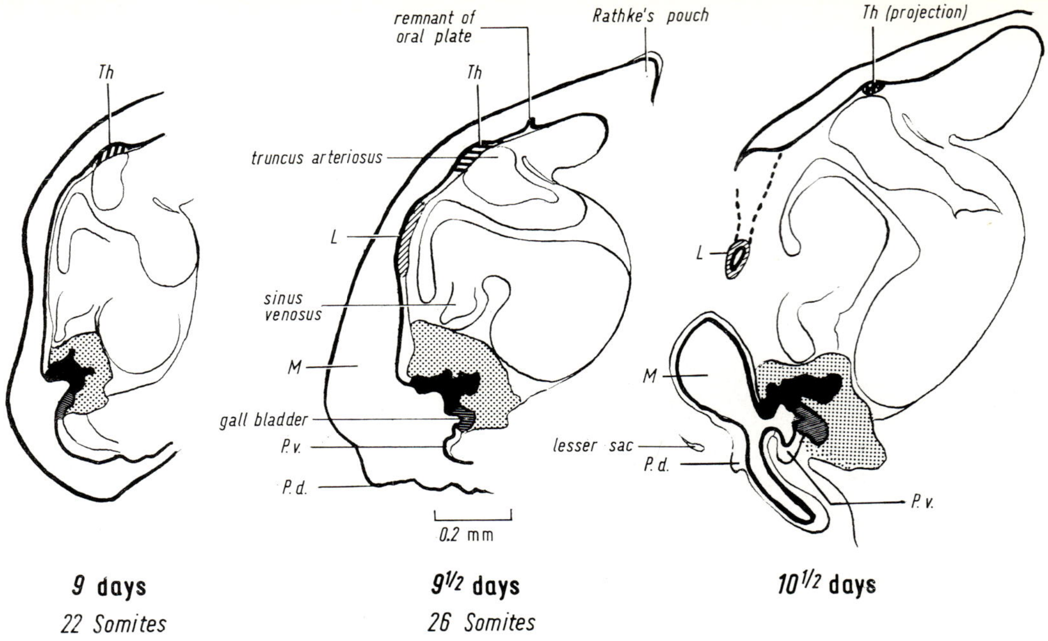

Fig. 123
Development of the intestinal tract, illustrated in 3 sagittal sections, 9–10 1/2 days, drawn to scale. *Th* Primordium of thyroid, *L* primordium of lung, *M* primordium of stomach, *P.d.*, *P.v.* pancreas (dorsale, ventrale); Mesenchyme of septum transversum: stippled; Liver anlage: full black.
KT 936 9 days, KT 997 9 1/2 days, KT 999 10 1/2 days

Central Nervous System

The *posterior neuropore* begins to constrict but is still open in all embryos examined (Fig. 117), even in those with 26 or 28 somites (Fig. 126). Thus, the closure is definitely retarded in mice compared with human embryos of the same somite number. The completion of closure in mice has been observed as late as the 32-somite stage.

During this period the brain develops very rapidly. The *optic vesicles* have relatively large stalks. From the outer periphery of the vesicles, the cells of the *optic neural crest* spread into the surrounding mesenchyme (Fig. 120). They probably become pigment cells of the uvea. The other divisions of the cranial neural crest [161] have formed prior to this stage, when the anterior neuropore was still open.

The development of the cranial blood vessels parallels the extremely rapid growth of the brain. The central nervous system is being invested by an extensive vascular plexus.

The *lens placode* and the *olfactory placode* appear as distinct thickenings of the surface epithelium (Fig. 119).

The *otocyst* (ear vesicle) is completely detached from the surface at the 22–24 somite stage. It is one of the most reliable characteristics for determining the degree of development. For a while, the closing rim of the otocyst is stretched into a ductlike stalk by which the ear vesicle is attached to the surface (Fig. 125). Sometimes the process of detachment of the right and left otocyst is slightly asynchronous, but the differences observed are slight. The epithelium of the otocyst is highly specialized and is entirely different from the simple skin ectoderm. It consists of tall cells, and all mitoses are localized in the superficial layer bordering the lumen. In this respect, it resembles the wall of the neural tube.

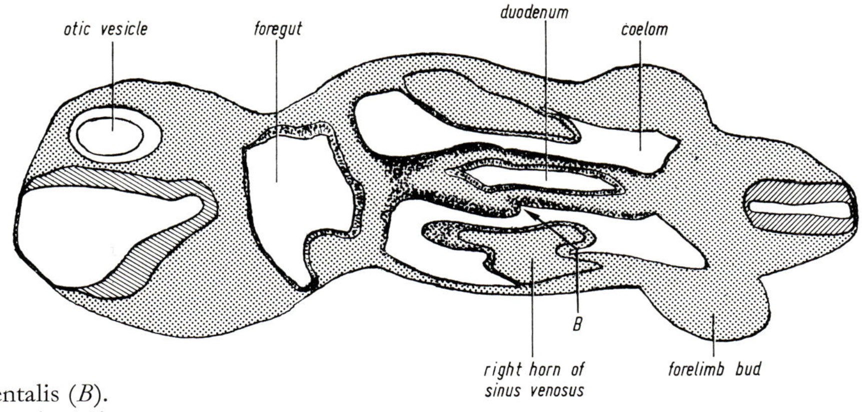

Fig. 124
Anlage of bursa omentalis (*B*).
KT 935/9, 9 days 3 h, 23 somites

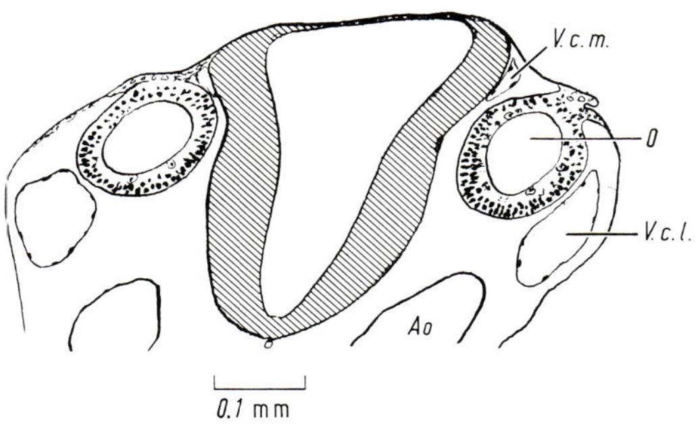

Fig. 125
Otic vesicle separating from epidermis, 9 days, 22 somites.
O Otic vesicle, *Ao* aorta, *V.c.l.* vena capitis lateralis, *V.c.m.* vena capitis medialis.
KT 935

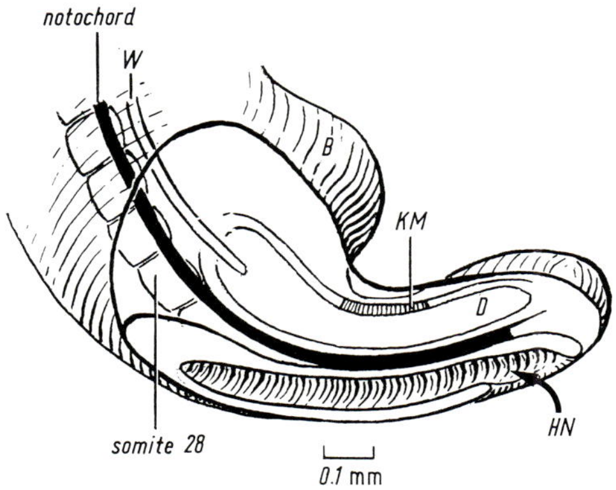

Fig. 126
Posterior part of the body. Graphic reconstruction.
10-day embryo, 28 somites; *HN* posterior neuroporus (*arrow*); *W* Wolffian duct; *KM* cloacal membrane; *D* hindgut, extending into tail bud; *B* hindleg-bud.
KT 939/3

Material	Age	
KT 935–37	9 days 3 h	4 with 15–20 somites (mentioned before) 3 with 22 somites 1 with 23 somites 1 with 24 somites
KT 997–98	9 days 10 h	1 with 17 somites. Posterior neuropore open 1 with 22 somites. Posterior neuropore open 1 with 23 somites. Posterior neuropore open 1 with 24 somites. Posterior neuropore open 2 with 25 somites. Posterior neuropore open 2 with 26 somites. Posterior neuropore open
KT 648–49	9 days 23 h	7 with 22–23 somites. Posterior neuropore open. Two of them measured 3.2 and 3.5 mm (unfixed)
KT 938–40	10 days	9 with 25–34 somites, Five measured 3.2–3.9 mm (unfixed)

Stage 16 Closure of Posterior Neuropore; Hind Limb Bud and Tail Bud

10 Days, 30–34 Somites, 3.1–3.9 mm Length (fresh)

Horizon XIII
Lens plate
homo = 27–29 days,
4.5 mm

External Shape

Some typical features of stage 16 are represented in Fig. 134. The hind limb bud becomes visible as a distinct bulge at the level of the 23rd–28th somite. The tail rudiment appears as a short stump (Fig. 127). The surfaces of the third and fourth branchial bars are distinctly concave, in contrast to the second. In this way the formation of the cervical sinus is initiated. The *lens plate* is usually slightly indented (Figs. 129–131). This corresponds to the early horizon XIV of Streeter. Lens development seems to be slightly advanced in mouse compared to human embryos. The olfactory placode is clearly indented, and is also slightly advanced in development compared to the human embryos.

The *otocyst* is invariably closed. It is no longer spherical, but more or less pear-shaped in its dorsal segment (Fig. 127).

The *posterior neuropore* begins to close.

Length: In the unfixed state, the overall length varies between 3.3 and 3.9 mm.

Circulatory System

The cardinal veins may be clearly recognized in intact embryos (Fig. 129). The heart forms a prominent bulge in the vicinity of the branchial bars. The structure of the heart is essentially the same as in stage 17 for which a detailed reconstruction is represented in Fig. 148.

Intestinal Tract

The most conspicuous change in this age group is the formation of the lung anlage (Fig. 132): The laryngo-tracheal groove branches from the oesophageal gut and the two primary bronchi sprout out in a T-like manner (Figs. 132, 133). The development of the intermediate portion of the intestinal tract was shown in Fig. 123. The hindgut still has the same appearance as in Fig. 126.

Urogenital System

The nephrogenic cord is now widely separated from its origin in the somite stalks. In the pronephric region, anterior to the 12th somite, only small clusters of cells are visible, and there is no structural organization. At the level of the 12th somite, spherical groups of nephrogenic cells may be recognized. They are arranged as vesicles at the level of the 13th to the 15th somite (Fig. 135). Generally, two such vesicles will form in each segment. Maturation proceeds slowly: At the 32 somite stage, the nephrotome of the 20th somite is not much more developed than in the 26 somite stage (Fig. 135). In one 26 somite embryo, there is a peritoneal funnel. There are no mesonephric glomerula at this stage.

The *Wolffian duct* is recognizable in the younger specimen examined, and it extends from the 13th to the 26th somite. In the older specimen, it has already reached the cloaca. At the point of contact, numerous pycnotic nuclei may be seen.
Primordial germ cells [103] can be seen as large cells containing much alkaline phosphatase. Some are already situated within the genital ridge, on both sides of the dorsal mesenteric attachment.

Central Nervous System

In most cases the neural tube is now completely closed. Compared to the human embryo of similar somite number, closure is delayed in mice. As an example, specimen KT 939/4 has 32 somites, and its posterior neuropore has just closed.
The separation and differentiation of the *otic vesicle* behaves nearly exactly the same in mice as in humans. The vesicle is now closed and completely separated from the epidermis. The endo-lymphatic appendage is marked off as a dorsal recess from the main cavity of the vesicle, giving it a more elongated appearance.
The *olfactory placode* is considerably thickened and slightly indented.
The *ganglia* of the cranial nerves appear as distinct blastemal condensations. In transparent specimens, the large trigeminal ganglion can easily be recognized, situated just anterior to the pontine flexure of the brain tube (Fig. 127). Posteriorly, the spinal ganglia are differentiated from the neural crest.
In older specimens, the *lens plate* is slightly indented (Fig 130).

Placenta

The loosely structured chorionic plate is traversed by the allantoic vessels. The ectoplacental plate is transformed into the labyrinth, and its margins are turned inward (Fig. 136). Consequently, the transition zone of the two yolk sac layers is also turned inward. The yolk sac cavity will later communicate here with the interplacental cavities which develop secondarily as small clefts within the labyrinthine cell mass, after disappearance of the original ectoplacental cavity [33].
The zone of "giant cells" (Fig. 136 *A*) borders the decidua which has many multinucleate cells in this region. Toward the mesometrium, the deciduous cells penetrate the myometrium and start to form the "metrial gland" [25].

Material	Age	
KT 938–40	10 days	4 with 25, 26, 27 and 28 somites (listed previously) 1 with 32 somites. Posterior neuropore just closed 1 with 32 somites. Posterior neuropore still open 1 with 34 somites. Posterior neuropore closed

Fig. 127 Embryo of 32 somites, from the right. Formalin fixed.
The trigeminal ganglion (*V*) is visible through the skin as well as the slightly pointed otic vesicle (*O*). Dashed line: level of the section Fig. 130. 13.5:1

Fig. 128 Embryo of 30 somites, on millimeter scale. Bouin fixed.
Nominal age 11 days, developmental age 10 days.
KT 945. 13.5:1

Fig. 129 Embryo of 33 somites, from the left.
Explanations see Fig. 134. 13.5:1

Fig. 130 Horizontal section through eye primordium, low magnification. Plane of section indicated in Fig. 127.
D diencephalon (3rd ventricle), *Tel* cerebral hemisphere.
KT 939/4, 32 somites. 100:1

Fig. 131 Detail of Fig. 130.
L Lens invagination, *R* retinal sheet. 270:1

Fig. 132 Cross section, level of lung primordium. The apparent asymmetry of the lung-bud is mainly due to the oblique plane of section.
V.c.a. Vena cardinalis anterior, *Ao* aorta dorsalis (paired).
KT 939/4, 10 days, 32 somites. 100:1

Fig. 133 Detail of Fig. 132.
L Laryngo-tracheal groove. 270:1

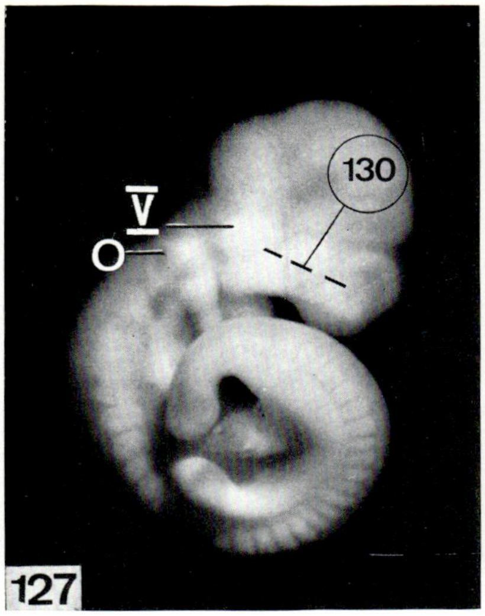

V
O
130
127

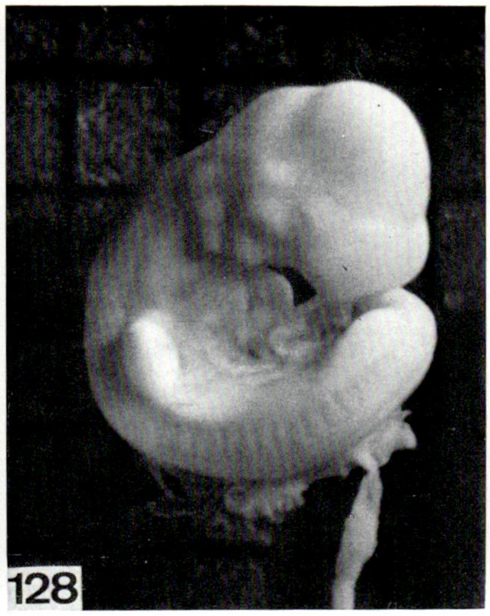

128

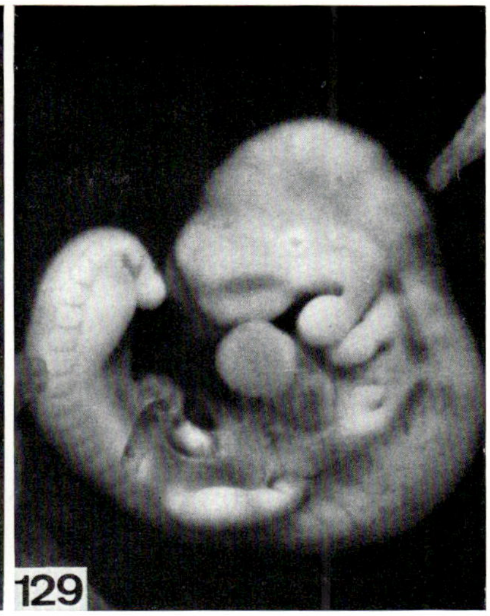

129

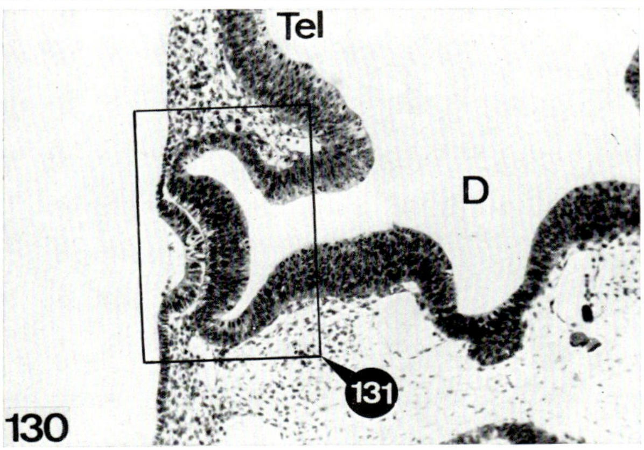

Tel
D
131
130

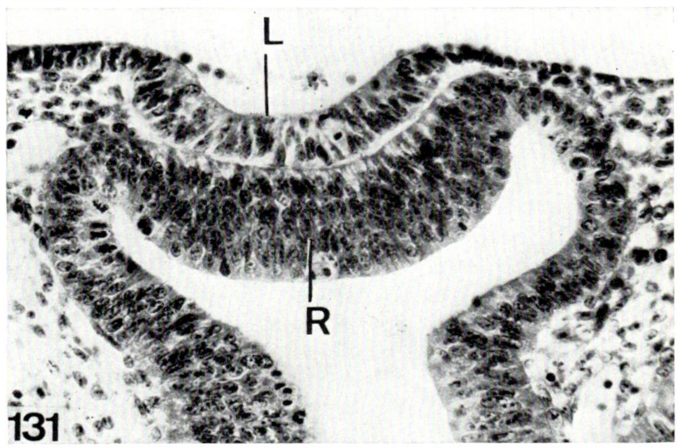

L
R
131

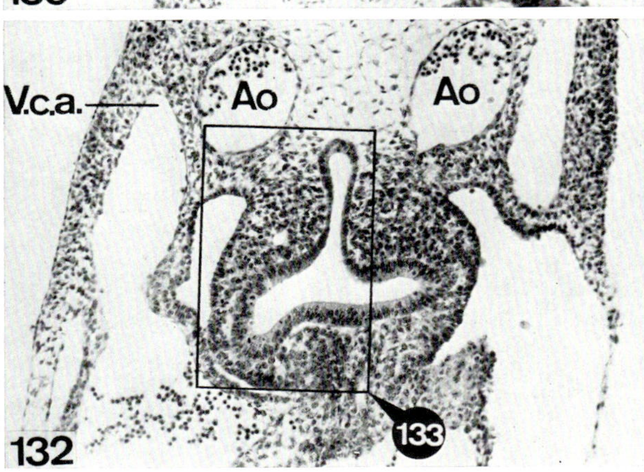

V.c.a.
Ao
Ao
133
132

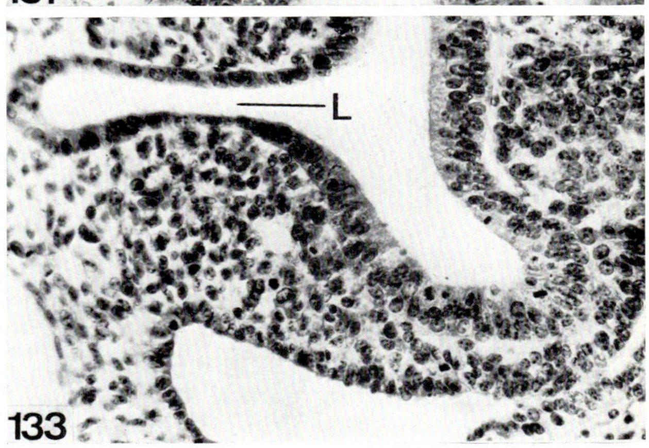

L
133

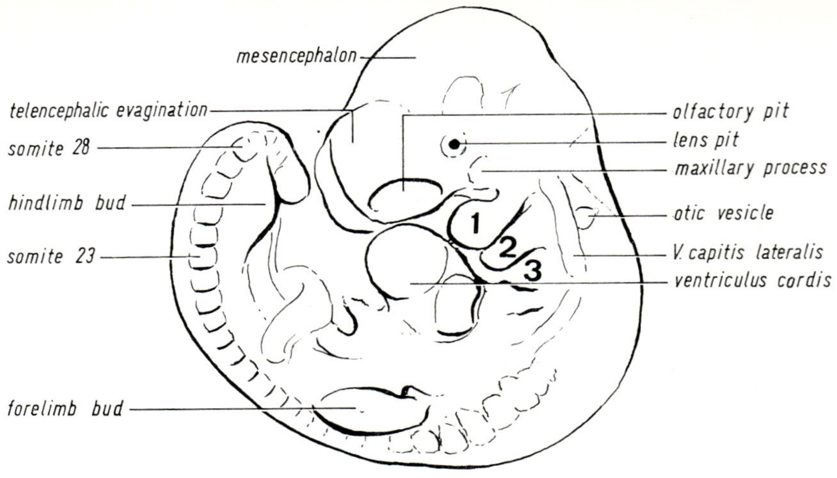

Fig. 134
Externally visible features of embryo Fig. 129, 33 somites

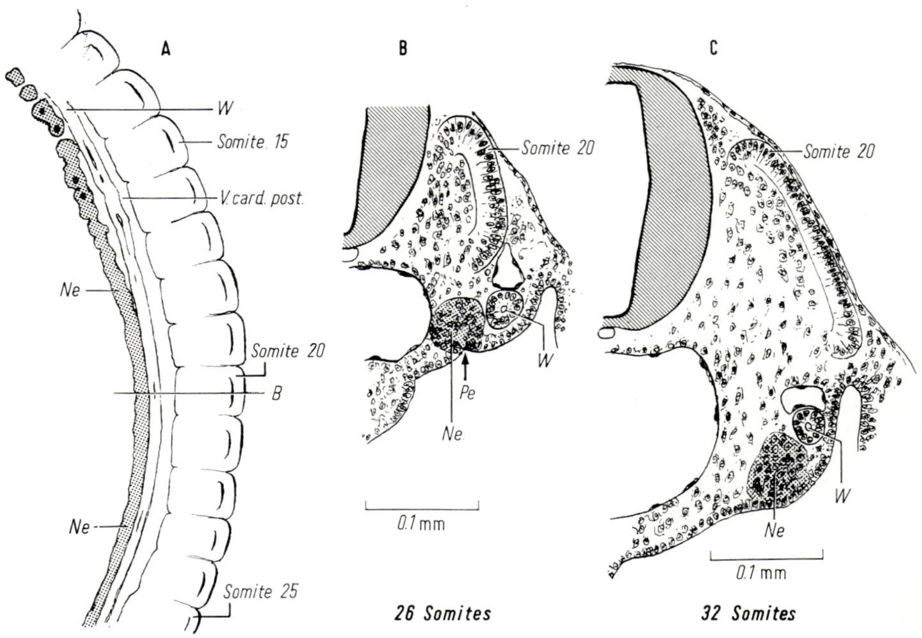

Fig. 135
Development of the mesonephros, 10 days.
A Drawing shows level of cross section; *B* cross section, embryo of 26 somites; *C* same cross section, embryo of 32 somites; *Pe* peritoneal funnel; *Ne* nephrogenic cord (dotted area); *W* Wollfian duct.
KT 939/2

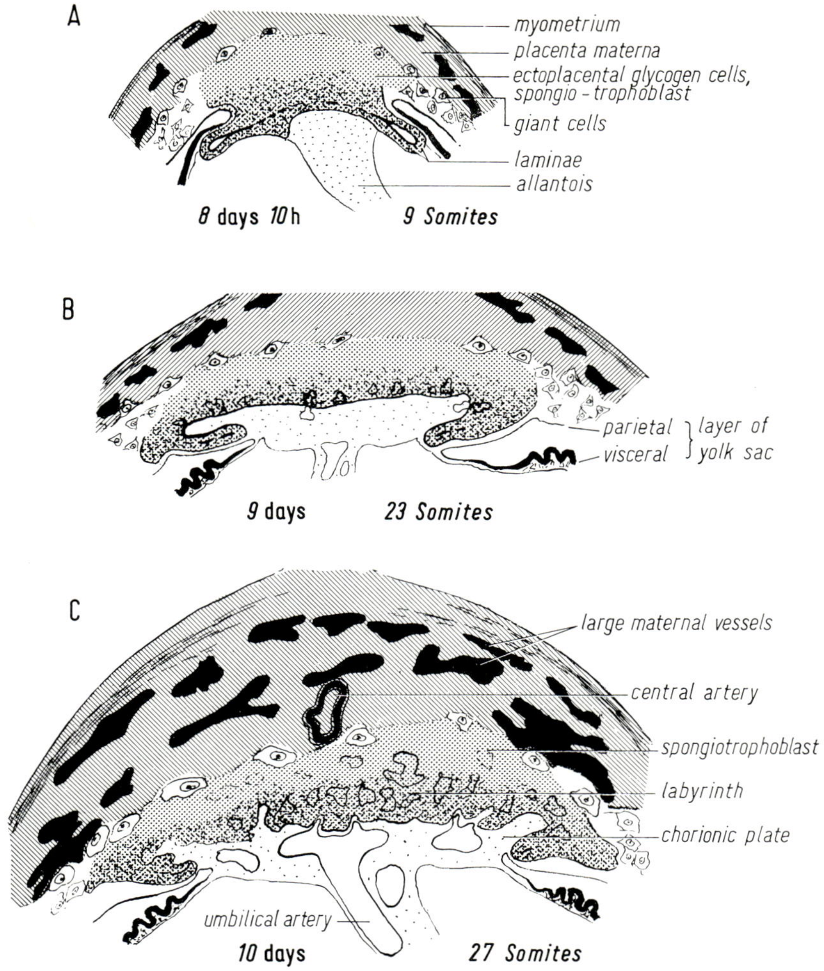

Fig. 136
Development of the placenta.
A Disappearance of the ectoplacental cleft (broken line between laminae), *B* margins of laminae turned inwards, *C* formation of labyrinth (dashed area = maternal tissue).
A KT 1001/4, *B* KT 935/9, *C* KT 938/4

Horizon XIV
homo = 32 days,
5–7 mm

Stage 17 Deep Lens Indentation

10 1/2 Days, 35–39 Somites, 3.5–4.9 mm (fresh)

External Shape

The *extremities* and *tail* are enlarging rapidly. They are no longer simple ridges and a short stump. Instead, they are projecting appendages and curve forward or inward. The tail is considerably longer now than in the human of this stage.

The *olfactory discs* are also more advanced, forming a distinct marginal lip. The *lens vesicles* form deep pockets (in younger embryos) or pore-like openings (older) to the surface. This is one of the best characteristics to use for assigning an embryo to this stage.

The first branchial bar is divided into large conspicuous maxillary and mandibular processes. The most cranial *somites* are becoming more and more indistinct. On the other hand, the most cranial *spinal ganglia* are easily recognized in transparent specimens. They are still completely lacking in the tail, where only somites are visible.

Behind the cloacal membrane, a short *ventral ectodermal ridge* (Grüneberg [195]) is recognizable (Fig. 146).

Length: The overall length of unfixed embryos varies from 3.5 to 4.9 mm.

Sagittal section (Fig. 145):

The most prominent feature is again the advanced development of the *brain tube*. The *optic recess* (groove) is a conspicuous depression, situated just posterior to the thick commissural field and anterior to the thin optic chiasma. *Rathke's pouch* is narrowed and the median thyroid is deepened. It is shown at higher magnification in Fig. 141.

Circulatory System

With the development of the yolk sac and the placenta, two complete *extraembryonic circuits* have been established: the yolk sac and the placental circulation. The yolk sac vessels, passing along the obliterated yolk sac stalk (Fig. 137), separate distally. Only in the region close to the embryo does the vitelline vein course parallel to the vitelline artery. It separates from it distally and independently proceeds to the yolk sac (Fig. 147).

The *aorta abdominalis*, now a single unpaired vessel, forks into two branches at the level of the hind limb buds. These branches are the two umbilical arteries and the left one is larger than the right one. The actual continuation of the original dorsal aortae is represented by two slender vessels proceeding in the dorsal body wall into the tail, nearly to its tip.

Both *umbilical arteries* unite ventral to the gut tube at the base of the umbilical ring to form an unpaired vessel.

The (unpaired) *vitelline artery* branches off from the aorta as a conspicuous vessel in the middle region of the trunk [49]. It passes cranially beside the head to the yolk sac (Fig. 147).

Both umbilical veins unite in the region of the umbilical ring by anastomoses, caudal to the umbilical and vitelline arteries. Distally, a large venous stem branches within the placenta into smaller vessels. Proximally, however, there are two asymmetrical umbilical veins which pass within the left and right body walls. The right umbilical vein is more than twice as large as the left.

The *heart* is still a single, curved tube. It has two distinct constrictions: the sulcus atrioventricularis and the sulcus sinu-atrialis (Fig. 148). The termination of the above mentioned vitelline and umbilical veins is reconstructed in Fig. 147.

Aortic arches: The first and the second arch are greatly reduced compared to the previous stage. The 6th is developing and sends a branch to the lung primordium.

Intestinal Tract

In the midline of the *pharyngeal region* the *thyroid rudiment* is represented by a narrow diverticulum (Fig. 141). Laterally, all *pharyngeal pouches* may be seen.

The *lung bud* is rapidly elongating. The left primary bronchus is shown in Fig. 143 as a small tube.

The short *esophagus* continues into the large eccentric *stomach* anlage. This asymmetric development results in the formation of coelomic pockets, which are conspicuous in microscopic sections (Fig. 145).

In the region of the *hepatic diverticulum*, the evaginations of the ventral pancreas and of the gall bladder primordium are distinctly visible (Fig. 144). The hepatic cords are invading the mesenchyme of the septum transversum. Future erythropoietic cells seem to stem from this mesenchyme and now join these epithelial cords. The dorsal pancreas begins to constrict at its base. It grows both caudally and towards the left. In Fig. 144 (marked *P.d.*) the wall of the basal constriction is cut tangentially.

The *umbilical loop* is only slightly curved and is clearly visible in Fig. 137. The obliterated vitello-intestinal duct (yolk stalk) is attached to its apex. The vitelline vessels described previously pass through its walls to the yolk sac.

The *cloacal membrane* has not yet ruptured.

Urogenital System

The Wolffian duct has now reached the cloacal wall, where it ends blindly. The ureteric bud is forming. The mesonephric tubules are longer and more numerous than in the previous stage. However, there are no mesonephric glomeruli.

Central Nervous System

The cerebral vesicles appear as distinct bulges. The *rhombencephalic* portion of the brain is relatively large. It has an extremely thin transparent roof and a thick and wide floor with marked neuromeric elevations [162] (Fig. 143, upper left).

In the floor of the 3rd ventricle, the optic recess is distinctly visible. In the *roof* of the *diencephalon*, some pycnotic cells can be regularly observed [158]. Later, at 11 or 12 days, there is a pronounced degeneration of cells [166].

The *deep lens pocket* [181] (Fig. 140), which will develop into a pore-like opening of the lens vesicle, is a characteristic of this age group.
The *otic vesicle* has developed a short endolymphatic duct (Fig. 149).
The *olfactory plate* is broad, but relatively flat, with a distinct rim (Fig. 140).
The *ganglia of the branchial nerves* are now well formed. The dorsal ganglion of the glossopharyngeal nerve is visible in Fig. 142, below the otic vesicle, close to the anterior cardinal vein (later: anterior jugular vein).
In the trunk region, the most anterior spinal ganglia may be seen lateral to the somites in transparent specimens.

Material	Age	
KT 909	10 days	4 embryos (all well developed), 3.3–4.4 mm
KT 944–46	10 days 23 h	5 embryos + 1 resorption, 3.5–4.5 mm, 33–35 somites
KT 999–1000	10 days 9 1/2 h	3 embryos + 2 resorptions, 4.0–4.5 mm

Fig. 137 Embryo of about 36 somites.
N Umbilical loop with projecting yolk sac stalk (obliterated). 11:1

Fig. 138 Embryo, formalin fixed, nominal age 10 days, but actually further developed.
Explanations in drawing Fig. 139.
KT 909. 20:1

Fig. 139 Drawing of embryo (Fig. 138).
Tel Cerebral hemisphere, *L* lens invagination, *Hl* hindlimb-bud, *Aa* forelimb-bud, *O* otic vesicle, *Ri* olfactory pit, *He* heart, *So somite* 4, *C2* ganglion cervicale 2. Branchial bars 1, 2.
KT 909

Fig. 140 Cross section through eye and olfactory pit.
L Lens pocket, *Ri* olfactory placode, *3.V.* 3rd ventricle.
KT 909. 100:1

Fig. 141 Thyroid primordium, sagittal section.
Th Thyroid pocket, *T.a.* truncus arteriosus.
KT 999, 10 days 9 1/2 h. 350:1

Fig. 142 Otic vesicle (*O*), oblique section (transverse-frontal).
Rh Rhombencephalon.
KT 909. 130:1

Fig. 143 Paramedian section.
Lu Lung bud (left stem bronchus), *Ao* aorta dorsalis, *S.v.* sinus venosus, *Coe* coelomic pockets, * = stomach. *Arrow:* neural crest.
KT 999, 10 days 9 1/2 h. 40:1

Fig. 144 Detail of Fig. 142.
Lz Hepatic cell cords, *G* anlage of gall bladder, *P.v.* pancreas ventrale, *P.d.* pancreas dorsale (tangential section). 270:1

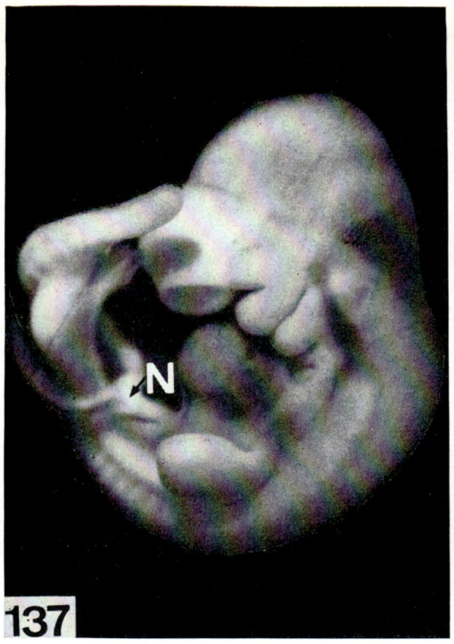

137

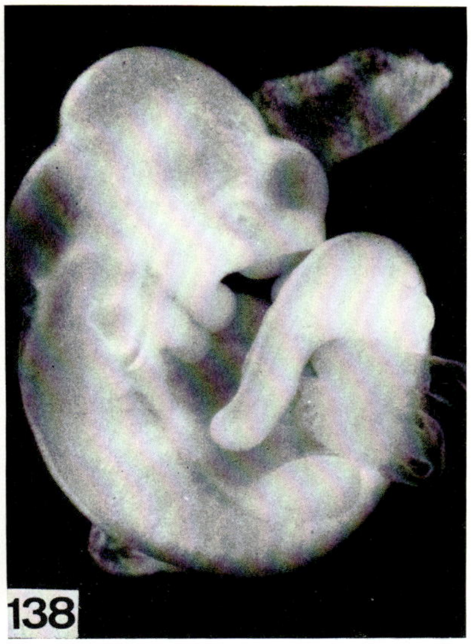

138

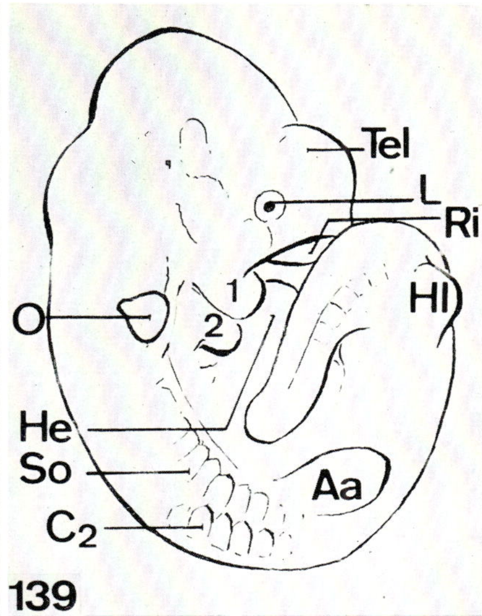

139

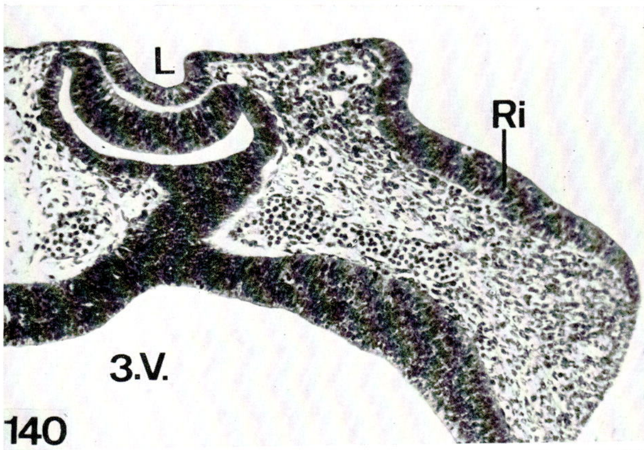

140

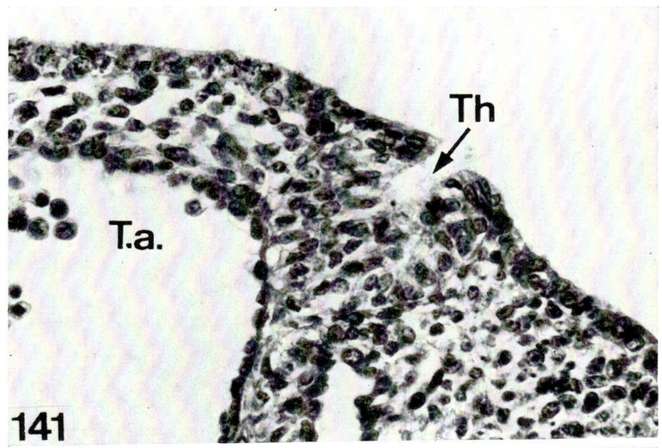

141

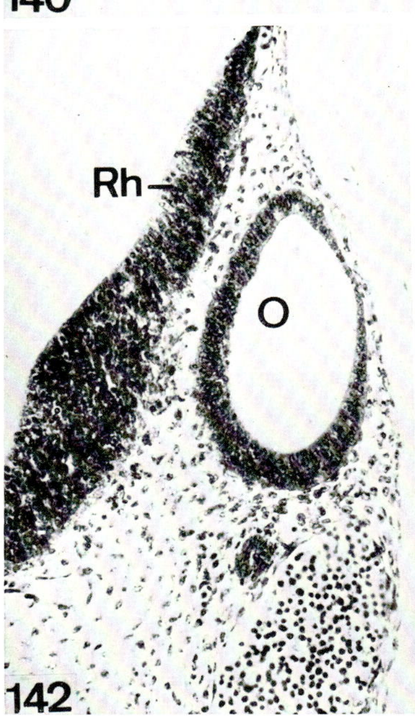

142

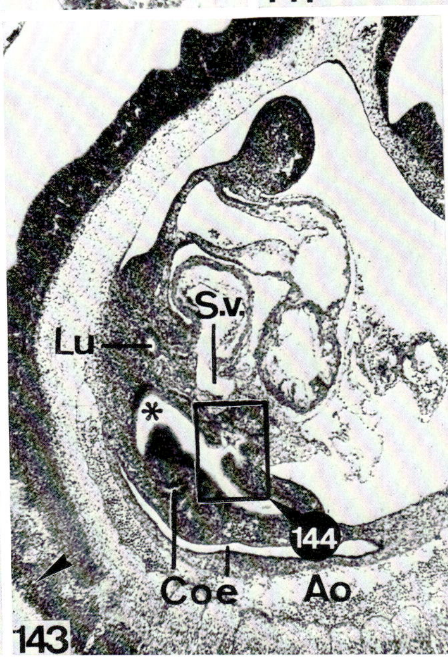

143

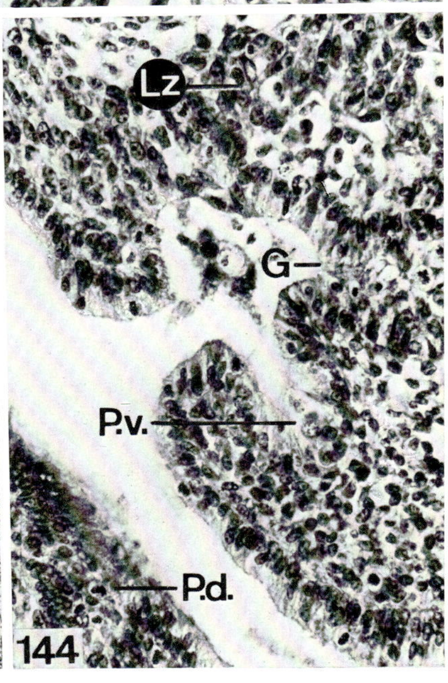

144

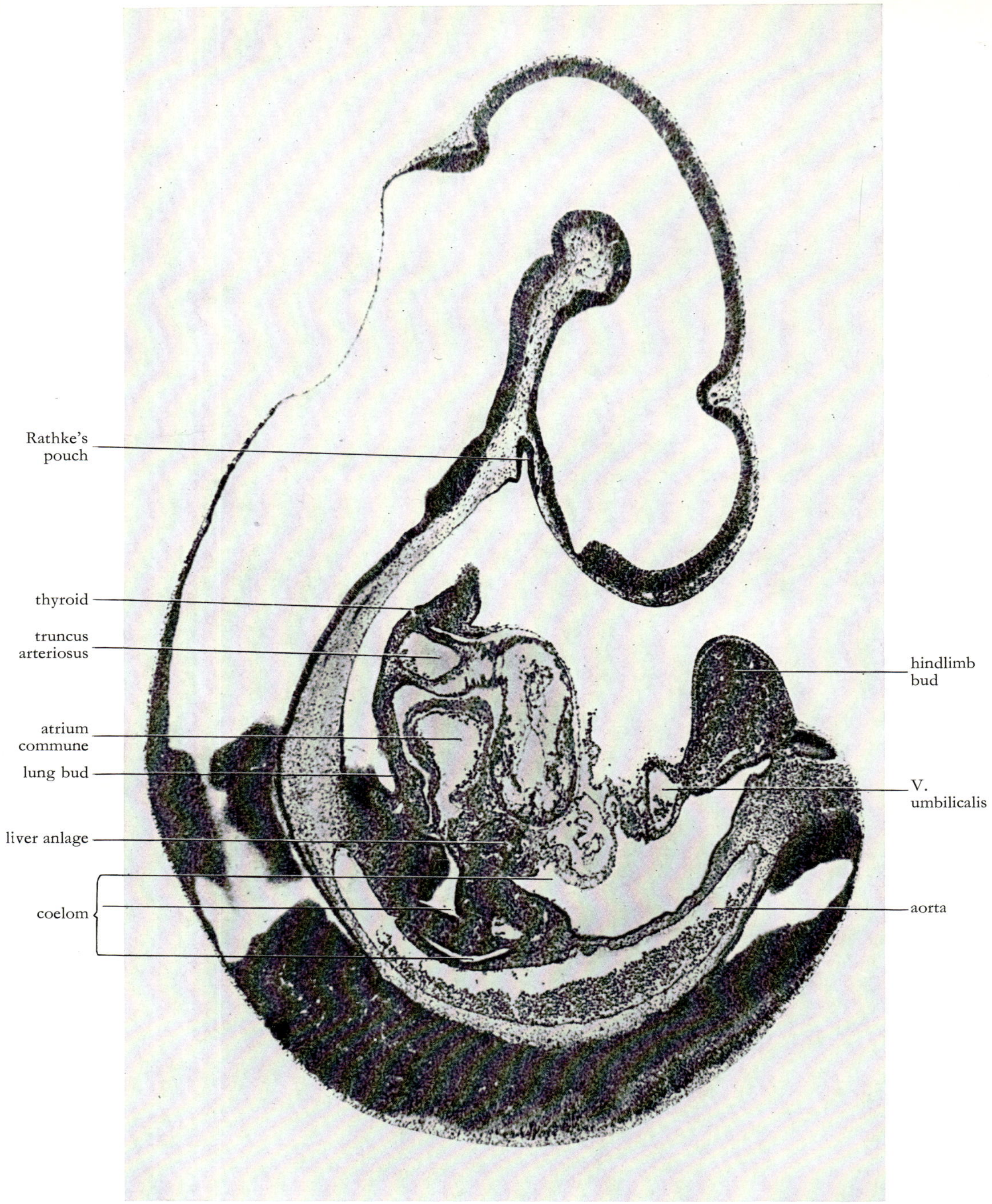

Fig. 145
Sagittal section, 10 days 9 1/2 h.
KT 999

Fig. 146
Proximal tail in cross section.
KT 909

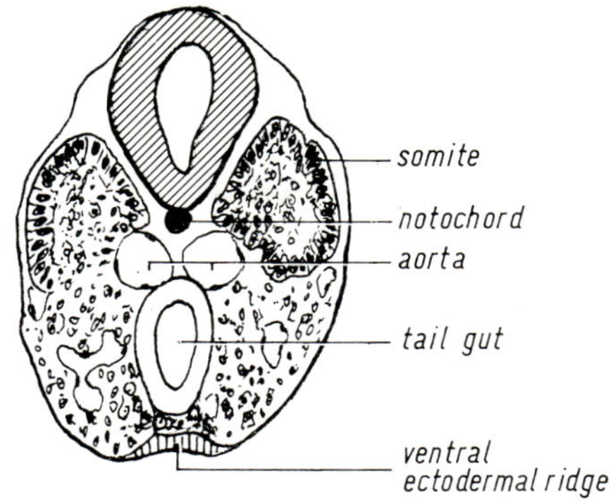

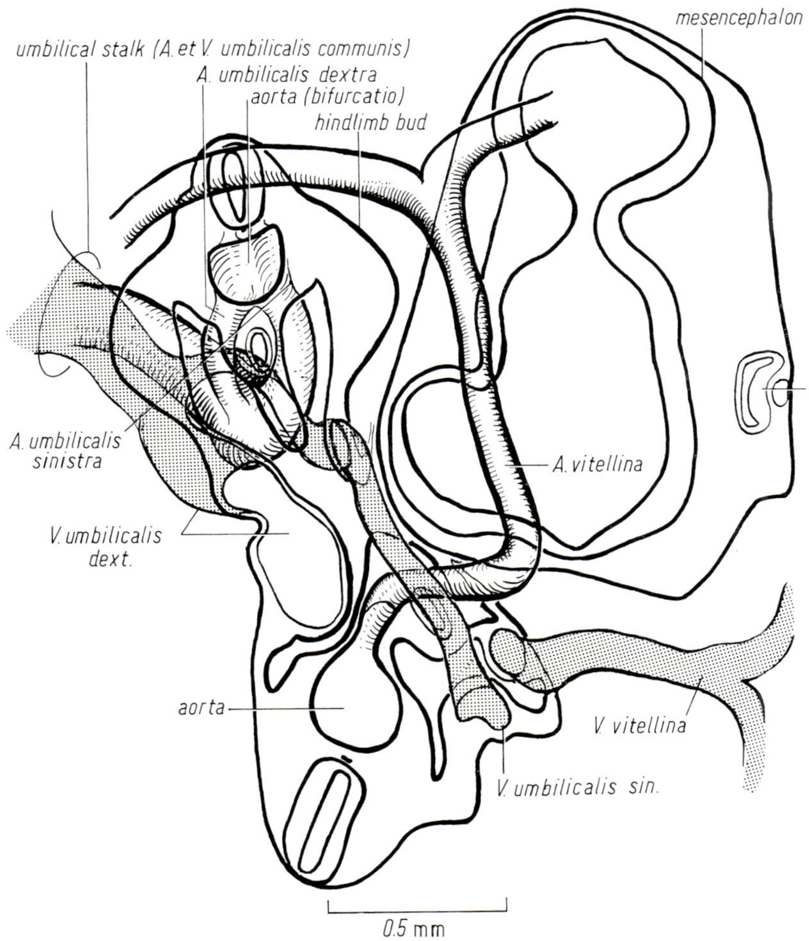

Fig. 147
Vascular connection with yolk sac and placenta. Reconstruction, starting from cross section through embryo.
KT 909

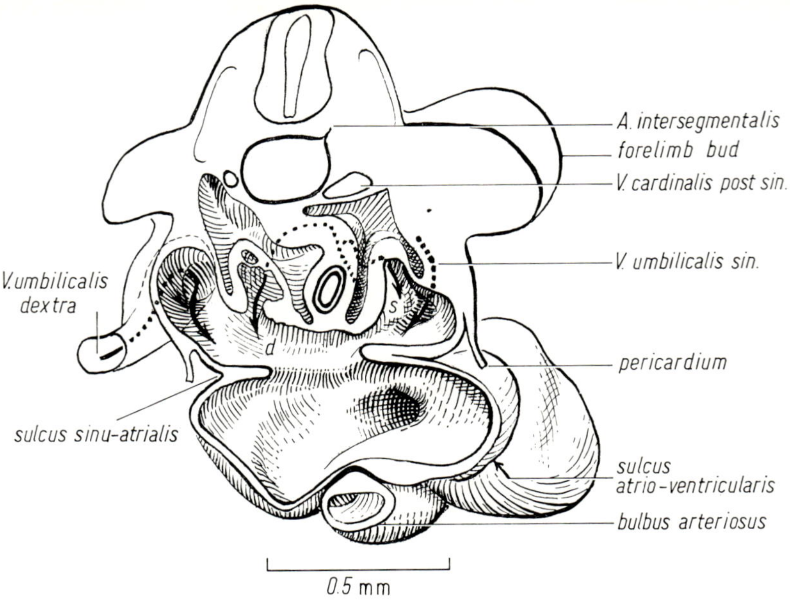

Fig. 148
Sinus venosus, cross section. Reconstruction.
View from cranial, *d* vena vitellina dextra, *s* vena vitellina sinistra.
KT 909

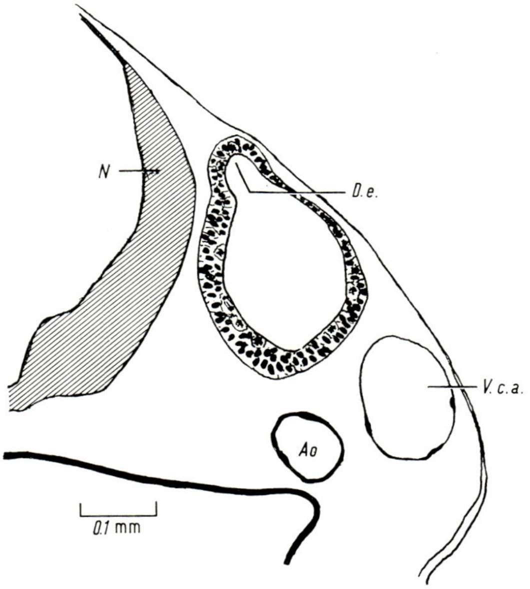

Fig. 149
Otic vesicle with endolymphatic duct (*D.e.*).
N Neural tube (rhombencephalon), *V. c. a.* vena cardinalis anterior, *Ao* aorta dorsalis.
KT 909, about 36 somites

Horizon XIV–XV
homo = 32–33 days
6–9 mm

Stage 18 Closure of Lens Vesicle

11 Days, 40–44 Somites, 5–6 mm

External Form

The somites in the cervical region are no longer visible. They can still be recognized in the posterior part of the body, especially in the tail. The long tail passes lateral to the head, over the deep nasal pit. The external features are shown in Figs. 150–152.
STREETER did not designate a special stage to mark the closure of the lens vesicle in man. Therefore, our stage 18 is between his horizon XIV (open vesicle) and XV (closed vesicle).
Length: Unfixed embryos are 5–6 mm long.
Sagittal section: The rapid growth and maturation of the brain is striking (Fig. 158). The infundibular recess is distinct, and the opening into the lateral ventricle is beginning to form. The liver is growing rapidly.

Circulatory System

The *heart* is still a curved undivided tube. The bulbar ridges can easily be recognized in sagittal sections (Figs. 145, 158). For the following stages (11 1/2 and 12 days), the differentiation of the bulbus arteriosus is described in more detail.
The primitive olfactory artery (arteria cerebri anterior) branches off from the *arteria carotis interna* at the same time as the nasal pits are developing.

Intestinal Tract

The *thyroid primordium* (Fig. 155) is shaped as a vesicle with a very thick basal wall. The lumen of this vesicle is a remnant of the thyro-glossal duct. It does not correspond to a thyroid follicle. In some embryos of this stage, the vesicle has separated from the surface.
The *lung bud*, stomach, and gall bladder are cut tangentially in the sagittal section shown in Fig. 158.
The epithelial lining of the dorsal mesentery of the stomach is thickened and represents the anlage of the *spleen* (Fig. 159). The thickening is part of the anterior splanchnic mesodermal plate (GREEN [83]).
The *cloaca* (Fig. 161) is not yet subdivided. The cloacal septum becomes evident in the next stage (Fig. 176).

Urogenital Tract

The *mesonephros* has not changed much since the preceding stage. The genital ridge is more distinct (Fig. 160), but diagnosis of the sex is not yet possible.

Central Nervous System

The brain is now more clearly subdivided, especially the diencephalon [172] with its basal recesses. The epiphyseal evagination, however, has not yet appeared. A marginal layer has formed in the brain stem (Fig. 154). The trigeminal and stato-acoustic ganglia may again be identified in transparent specimens (Fig. 152).

The *otocyst* has an elongated endolymphatic duct.

The *lens vesicle* begins to detach from the ectoderm (Fig. 153). In 2 out of 13 embryos of this group, the lumen had just lost its communication with the amniotic cavity.

The *olfactory plate* is now considerably deepened (Figs. 153 and 162). The bordering rims are beginning to unite. In the human, this stage is reached at a much later developmental phase, in horizon XVI.

Vertebral Column

Sclerotomic fissures are appearing in the trunk region. Later, they become more distinct (Fig. 186).

Material	Age	
KT 949–51	11 days 5 h	8 embryos, 4.8–5.1 mm length. 5–7 tail-somites
KT 604–5	11 days	6 embryos, 5.0–6.2 mm length

Fig. 150 Embryo of 11 days 5 h, Bouin fixed, on millimeter scale.
KT 949. 13:1

Fig. 151 Embryo of 11 days 5 h, Formol fixed, 5.3 mm length.
Explanation in drawing Fig. 152.
KT 950. 12.5:1

Fig. 152 Explanation of Fig. 151.
M Mesencephalon, *S* (broken line) plane of section Fig. 153, *Tel* hemisphere, *G5* trigeminal ganglion, *G7–8* ganglia of 7th and 8th cranial nerves, *O* otic vesicle, *Aa* forelimb bud, *Hl* hindlimb-bud, *1* mandibular process, *2* hyoid arch

Fig. 153 Eye anlage, frontal section, 11 days.
L Lens vesicle, in the process of closure; *A.c.* arteria carotis interna; *Hy* enlarged vessel of hyaloid plexus; *Ri* olfactory pit.
KT 605. 100:1

Fig. 154 Otic vesicle with endolymphatic duct (*De*).
Section parallel to "*S*" in drawing Fig. 152, same embryo of 11 days; *Rh* wall of rhombencephalon; *V.c.a.* vena cardinalis anterior.
KT 605. 100:1

Fig. 155 Thyroid primordium, sagittal section, 11 days 5 h.
Th Ductus thyreoglossus (closed here).
KT 949/6. 560:1

Fig. 156 Placenta with 11 days 5 h.
Section of whole uterine wall. Embryo in situ. *A* Amnion, *Au* arteria umbilicalis, *D* yolk sac, visceral layer.
KT 951/3. 18:1

Fig. 157 Detail of Fig. 156.
La Labyrinth of placenta, *eG* ectoplacental glycogen cells, *Rz* trophoblastic giant cells.
100:1

150

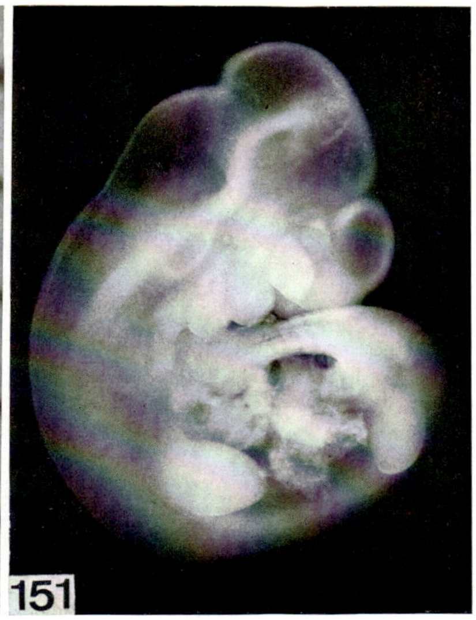
151

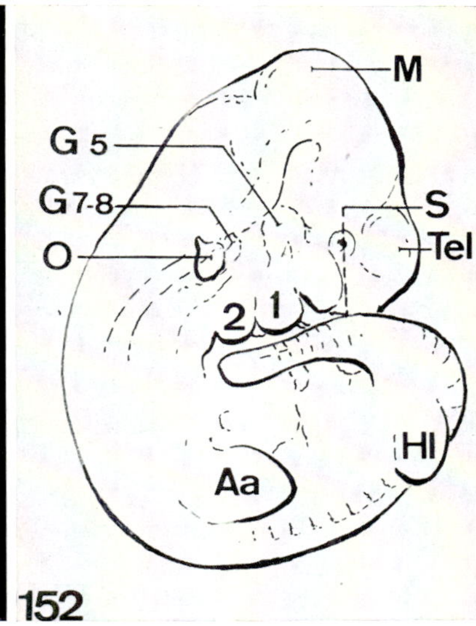

152

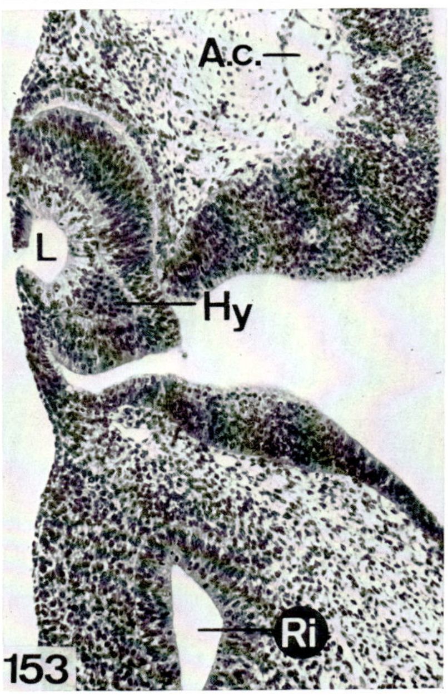

153

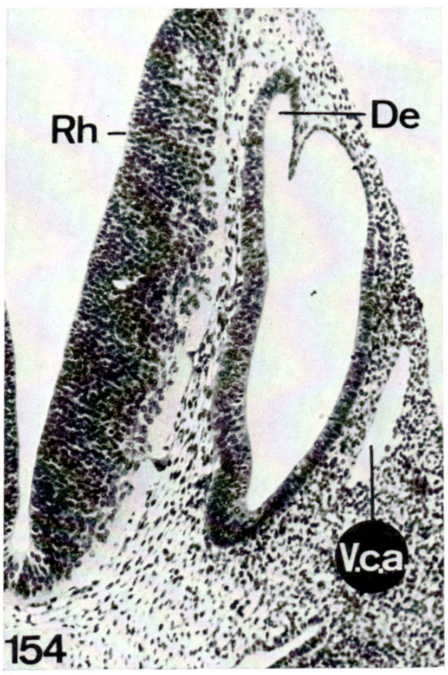

154

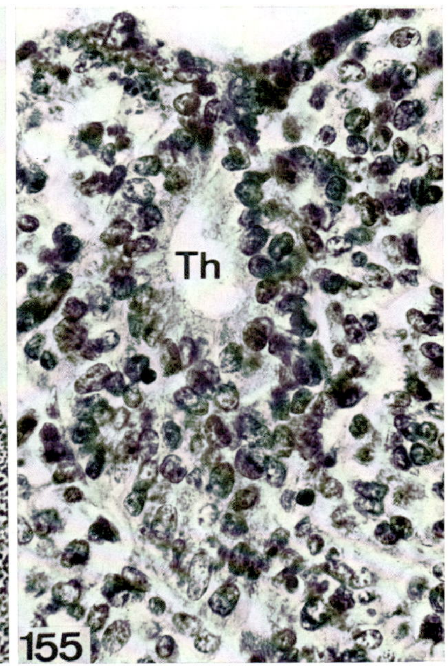

155

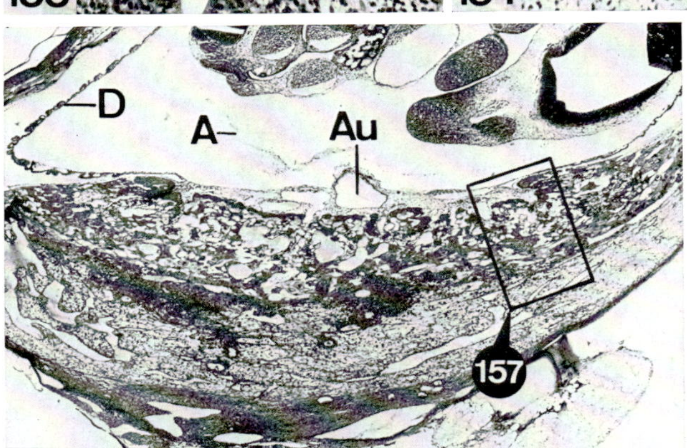

156

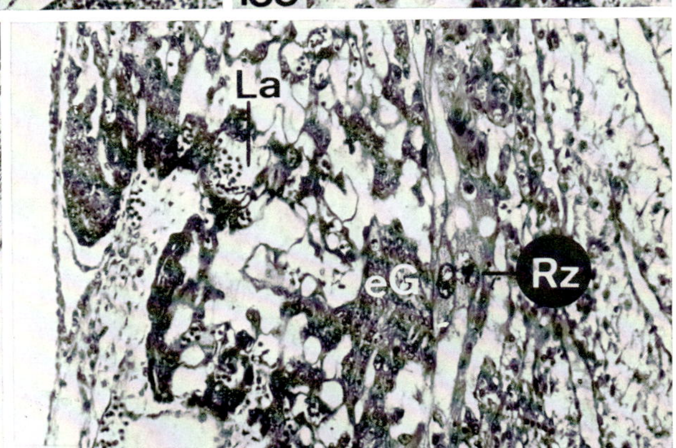

157

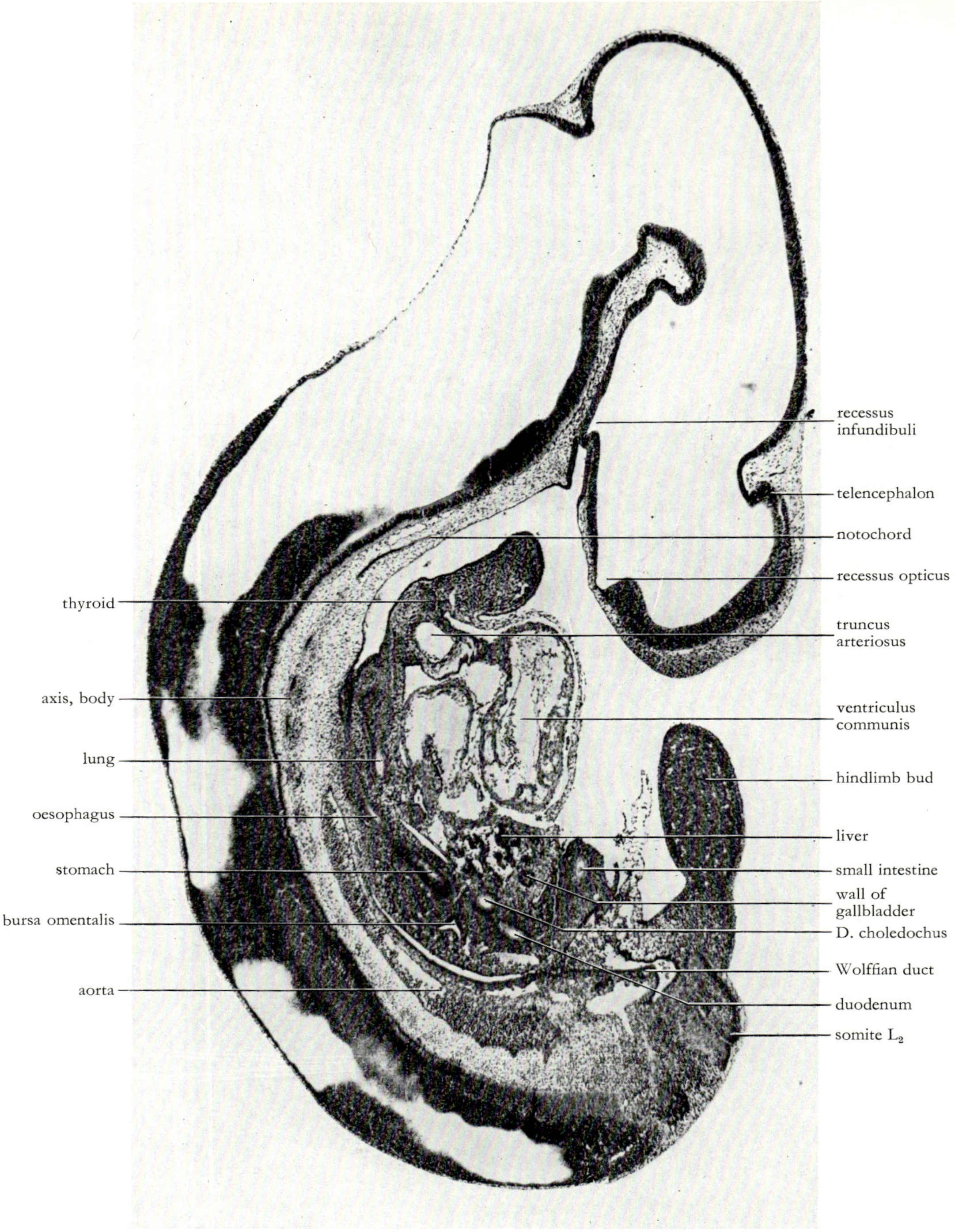

Fig. 158
Sagittal section, 11 days 5 h, 5 mm length.
KT 949/6

Fig. 159
Section through upper thoracic region and anterior limb-bud, 11 days 5 h.
KT 951/3

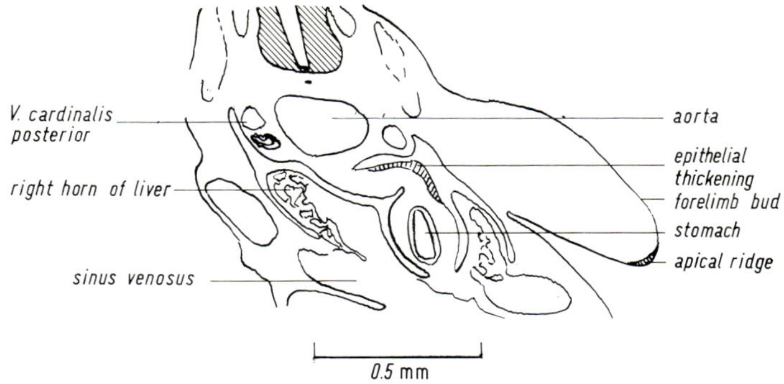

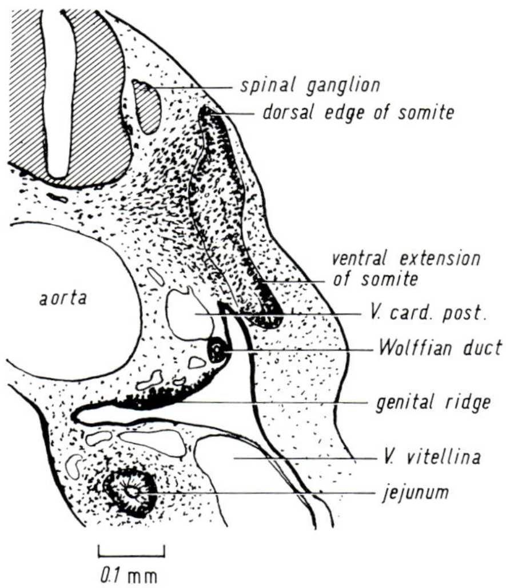

Fig. 160
Cross section through lower thoracic region, 11 days 5 h.
KT 951/3

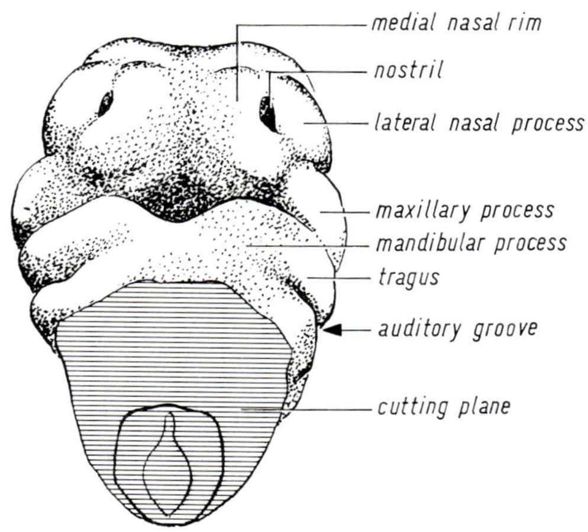

Fig. 162
Ventral view of the head, 11 days. Deep olfactory pit. Nasal folds fused posteriorly

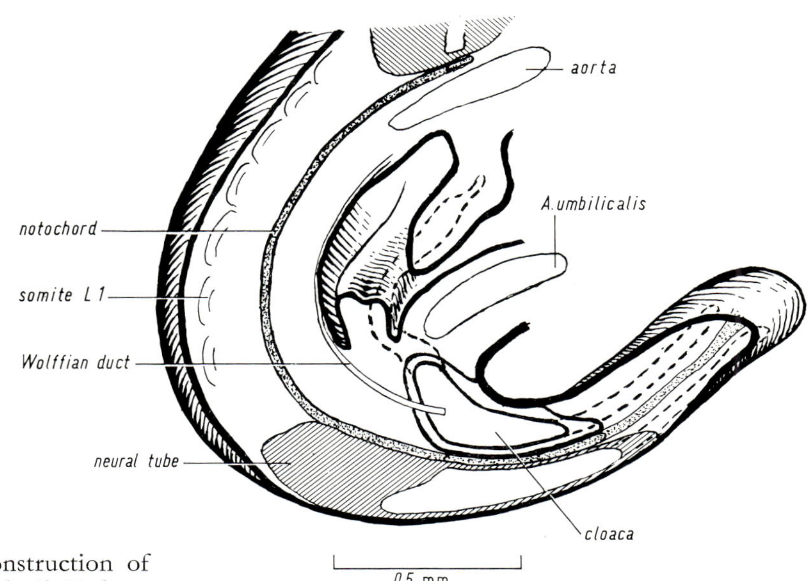

Fig. 161
Sagittal section and reconstruction of posterior trunk region and tail, 11 days.
KT 605/2

Stage 19 Lens Vesicle Completely Separated from Surface

11 1/2 Days, over 45 Somites, 6–7 mm

Horizon XVI
homo = 37 days
9–11 mm

External Form

A footplate ("handplate") has formed in the anterior limb bud, indicated by a definite constriction (*arrow* in Fig. 163). The posterior limb buds are not yet divided into leg and foot. In the human, both hand- and footplates are delimited in horizon XVI (STREETER). Their development is dissimilar to that in the mouse, and they are less useful to use as landmarks to indicate developmental stage.

The six low tubercles which will form the pinna can be discerned.

The nostrils are narrowed to small slits, and the nasolacrimal grooves are clearly visible.

The posterior somites are still sharply defined, and the tail is considerably longer than in the preceding stage.

Length: Most are 6–7 mm long. A few specimens of this age group may be somewhat smaller (KT 611).

Sagittal section (Fig. 171): Compared with the preceding stage, there is little change in the brain. The opening of Rathke's pouch is constricted. The thyroid primordium is growing deeper and is losing its lumen. The mesenchyme of the anterior primitive intervertebral discs is condensing. In the lumbar region and posterior to it, there is a gradual transition from loose mesenchyme to condensation of the posterior sclerotomic halves of the somites. The Atlas is visible in Fig. 175.

Circulatory System

Within the *arterial system*, the aortic arch complex has become more mature (Fig. 172). The most anterior arch has disappeared, and the second is reduced. The 3rd and 6th are well developed. All of them connect dorsally with the dorsal aorta. The vertebral artery is now forming parallel to the dorsal aorta. The primitive arteria cerebri anterior (olfactory artery) has enlarged considerably along with the further development of the nose.

In the *venous system*, the asymmetry of the umbilical veins is striking. The sinus venosus receives the cardinal veins, the right umbilical vein and the hepato-cardiac channel (= primitive inferior vena cava) (Fig. 173). The sinus venosus conducts the blood through a small sinu-atrial opening into the atrium.

The *atrium* is nearly completely partitioned off by the septum primum. The original broad communication is constricted to the foramen I, while the foramen II is forming above (Fig. 174).

The *ventricle* is still unpaired (= ventriculus communis). It is delimited from the atrium by the atrio-ventricular cushions (Fig. 175). This figure illustrates how the arterial outlet is incompletely subdivided by the developing bulbar ridges. It also shows how the pulmonary artery branches off as a thin vessel from the 6th aortic arch. The opening of the 6th aortic arch into the large dorsal aorta is also visible in Fig. 175.

Intestinal Tract

In the region of the *foregut*, the elevation of the tongue is not yet apparent. Anterior to the foramen caecum, there is a small indentation (Fig. 176), which must not be confused with it. The thyroid primordium has invaded more deeply into the underlying mesenchyme (Figs. 171, 176), and the primary (= stem) bronchi are developing secondary branches (= lobar bronchi). Lung lobation is determined genetically [61].

The pharyngeal pouches are now beginning to become specialized structures such as the thymus and parathyroid. They are not yet completely separated from the pharyngeal epithelium.

The *stomach* is much enlarged. It is separated from the outgrowing pancreas by the lesser sac (bursa omentalis). The *liver* is composed of broad hepatic cords, which are separated by large sinusoids containing nucleated erythrocytes. Hematopoietic foci are found intermingled with the hepatic cords.

The *caecum* is recognizable by a slight distension of the colon. It marks a clear distinction between colon and small intestine. Within the cloaca the uro-rectal septum is developing (Figs. 176, 177). The cloacal membrane has not yet ruptured. Posteriorly, the hindgut is continued by the tail-gut which terminates in the blastema of the tail tip. The tail-gut is very narrow, except for a slight terminal distension, and it has some pycnotic cells. The ventral ectodermal ridge [186] is confined to the region near the tail tip.

Urogenital System

The urogenital system is developing rapidly. The genital ridge, still in an indifferent state, contains numerous gonocytes (Fig. 168), which have now completed their migration [103]. The Wolffian ducts terminate blindly at the cloaca. The ureteric buds are considerably distended. They are surrounded by condensed metanephrogenic tissue [100] (Figs. 167, 177). The *mesonephros* is more mature than in the preceding stage. Several glomerulusanlagen may now be recognized, and the mesonephric tubules are more distinct (Fig. 178).

Central Nervous System

Eye: The most conspicuous feature of this age group is the complete closure of the lens vesicle and its detachment from the ectoderm (Fig. 166).

The pigment layer of the optic cup contains numerous cells with pigment granules.

Ear: The endolymphatic duct is longer, and the utricle and saccule are becoming discernible.

Olfactory organ: The nose-pit is separated from the oral cavity by the bucco-nasal membrane (Fig. 179). The epithelial wall of HOCHSTETTER is invaded and perforated by mesenchyme (Fig. 169). In the medial wall, the invaginated organ of JACOBSON is clearly visible.

The *ganglia* of the cranial nerves and the spinal ganglia are well developed and the anlage of the *sympathetic trunk* is discernible (Fig. 175).

The cellular degeneration which began at 10 1/2 days in the anterior part of the roof of the 3rd ventricle has progressed further.

Extraembryonic membranes: In the yolk sac, numerous well vascularized superficial folds have formed.

Material	Age	
KT 1004–6	11 days 10 h	6 embryos, 6.1–6.5 mm (unfixed) 1 resorption
KT 630–31	12 days	7 embryos, 6–7 mm (copulation age advanced)
KT 611–12	11 days 21 h	4 embryos, 5–5.5 mm 2 embryos, 2–2.5 mm (hypoplastic)

Fig. 163 Embryo, right side, on millimeter scale, 11 1/2 days (nt 114). *Arrow:* indentation delimiting the forefoot-plate. 9:1

Fig. 164 Embryo, left side, explanation in drawing Fig. 165 (nt 114). 9:1

Fig. 165 Explanation of Fig. 164.
Tel Cerebral hemisphere, *M* mesencephalon, *Ns* nostrils, *Rh* rhombencephalon, *Tr* nasolacrimal groove, *1*, *2* branchial bars, *V.c.a.* vena cardinalis anterior, *He* heart (ventricle), *So* somite (1rst lumbar), *Aa* forelimb bud, *Hl* hindlimb bud, L_3 spinal ganglion (3rd lumbar), *L* lens vesicle (just closed)

Fig. 166 Eye, frontal section.
A.c. Arteria cerebri anterior, *Pi* pigment epithelium, *R* retinal layer, *Hy* hyaloid plexus, *L* lens vesicle (just closed).
KT 630a/6. 130:1

Fig. 167 Caudal end of Wolffian duct, with ureteric bud.
Coe Coelom, *Ur* ureteric bud, *W* Wolffian duct, *Me* metanephrogenic tissue.
KT 630b/3. 130:1

Fig. 168 Sagittal section through germinal ridge.
Coe Coelom, *Ge* gonocyte.
KT 630b/6. 700:1

Fig. 169 Cross section through forebrain and olfactory apparatus.
A Anterior section, nostril; *B* intermediate section: dissolution of epithelial wall; *Tel* telencephalon; *Sc* tail, cross sectioned; *J* vomeronasal organ (Jacobson).
KT 630a/7. 40:1

Fig. 170 Cross section through forebrain and olfactory apparatus.
C Posterior section with bucconasal membrane, *Tr* nasolacrimal groove, *C.s.* corpus striatum.
KT 630a/7. 40:1

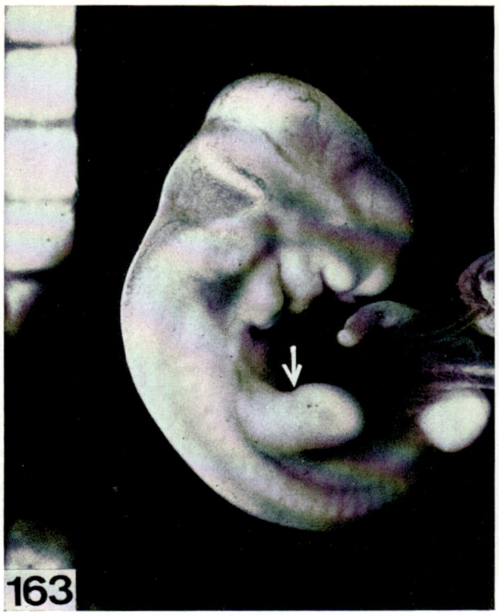
163

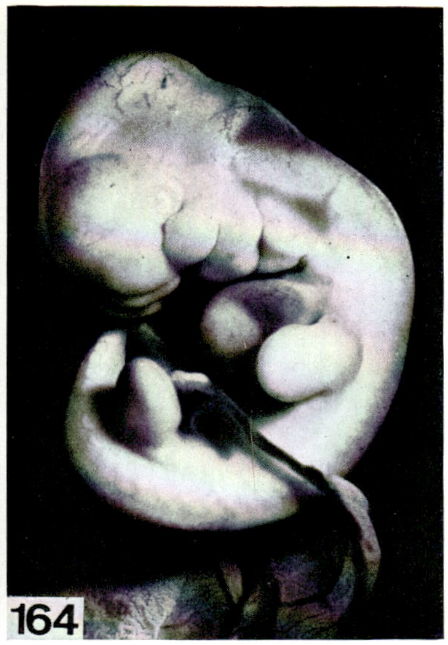
164

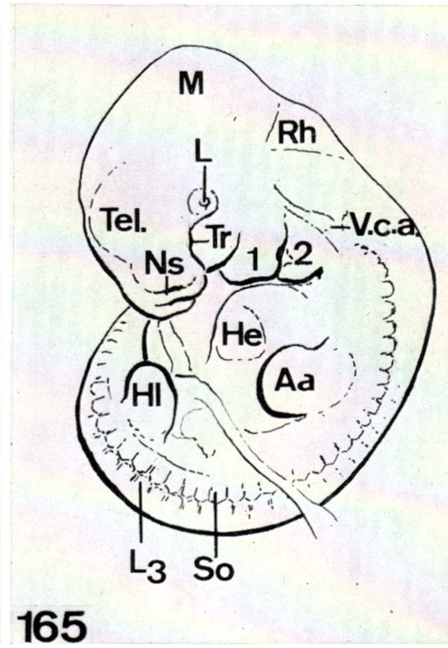

165

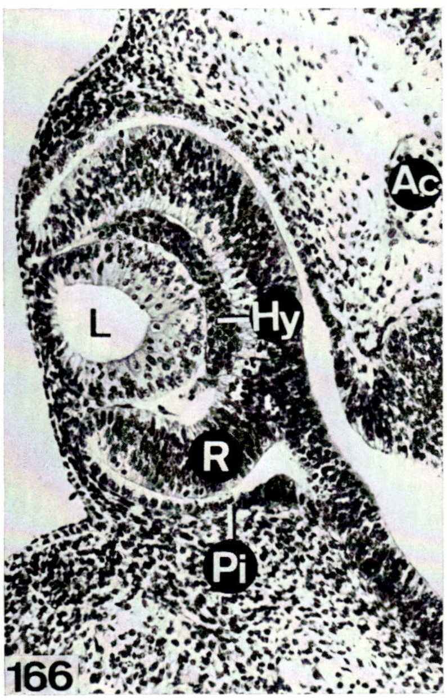

166

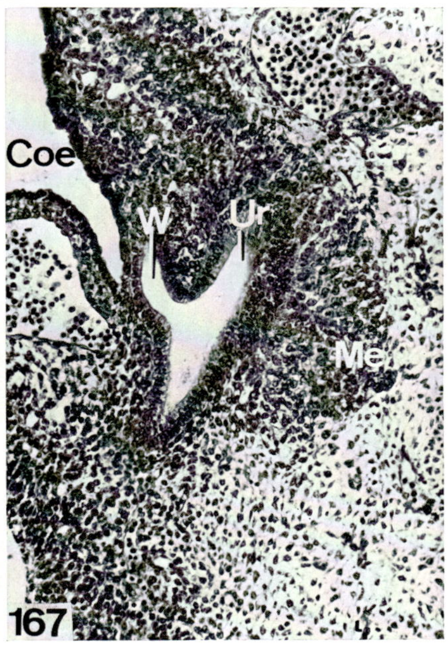

167

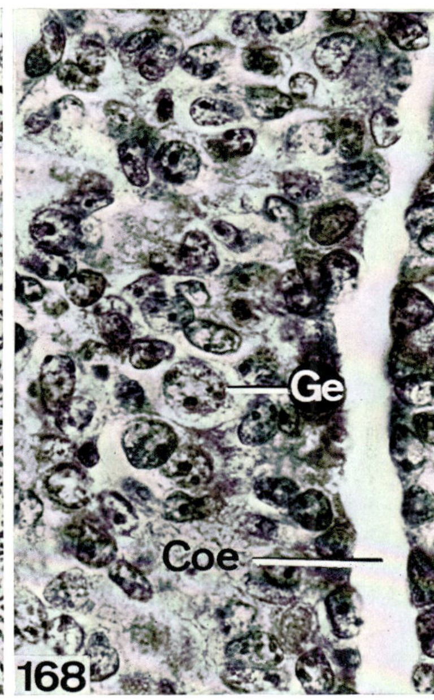

168

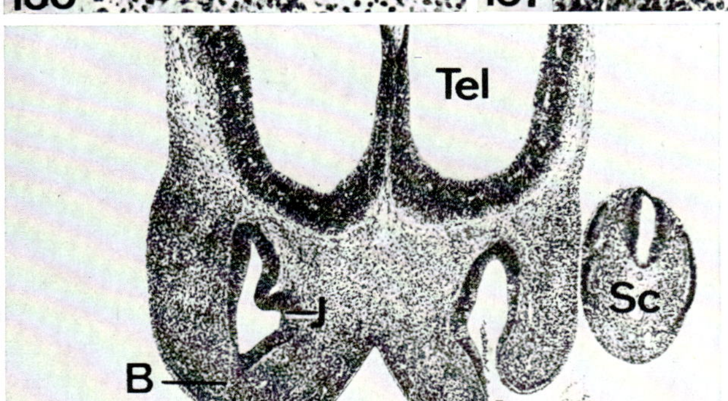

169

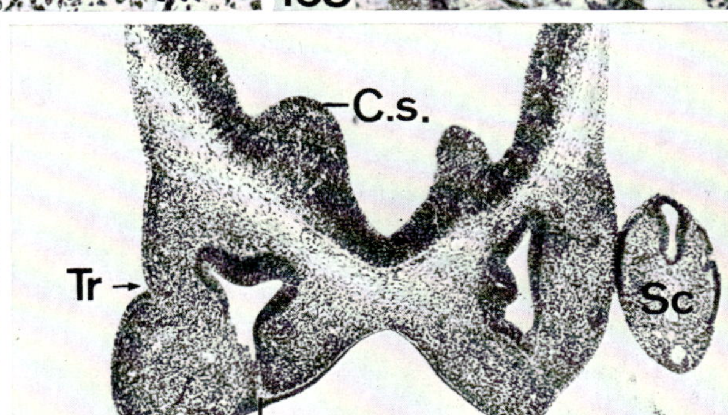

170

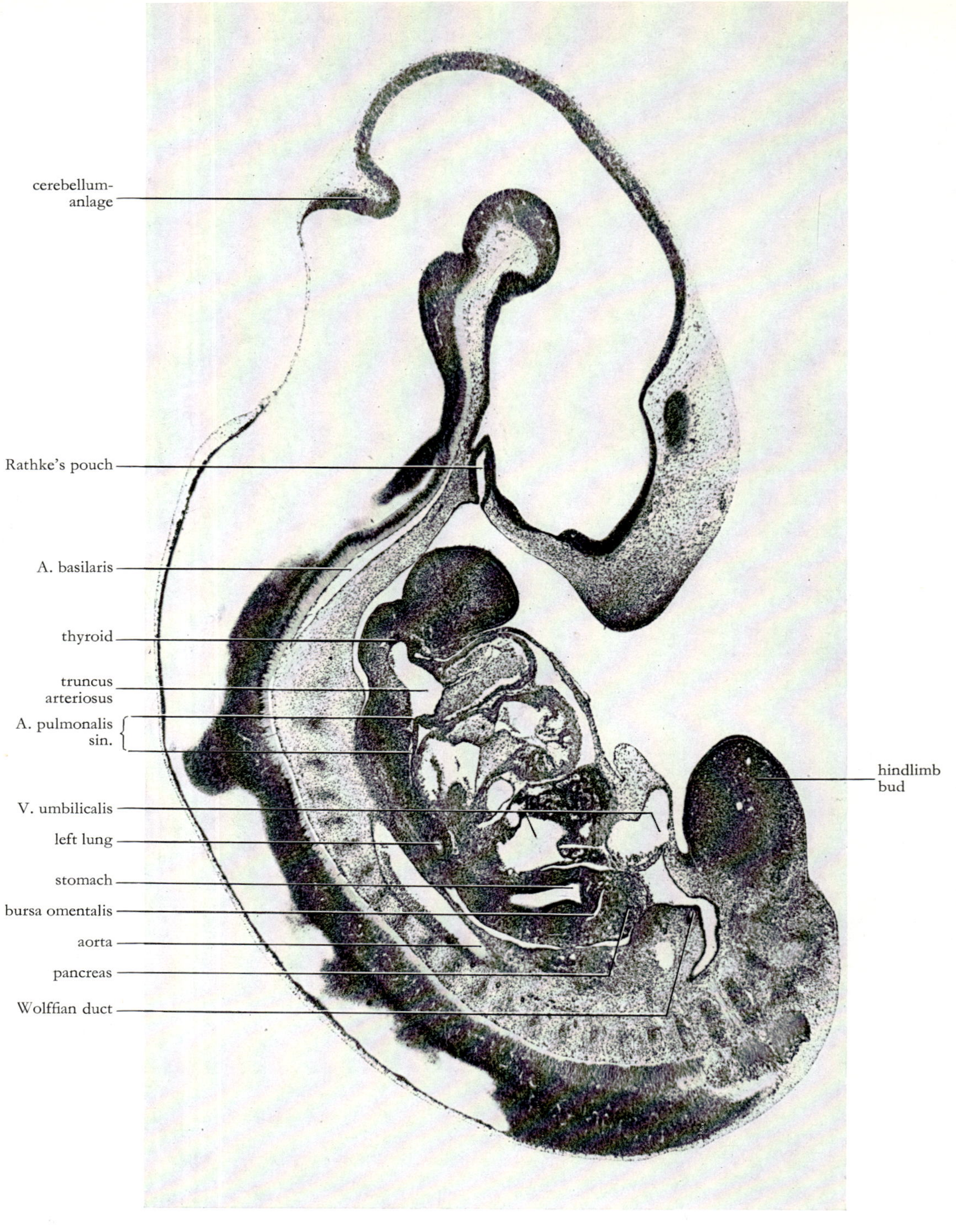

Fig. 171
Sagittal section, 11 days 10 h, 6.3 mm length.
KT 1005

Fig. 172
Reconstruction of the main blood vessels.
KT 630b

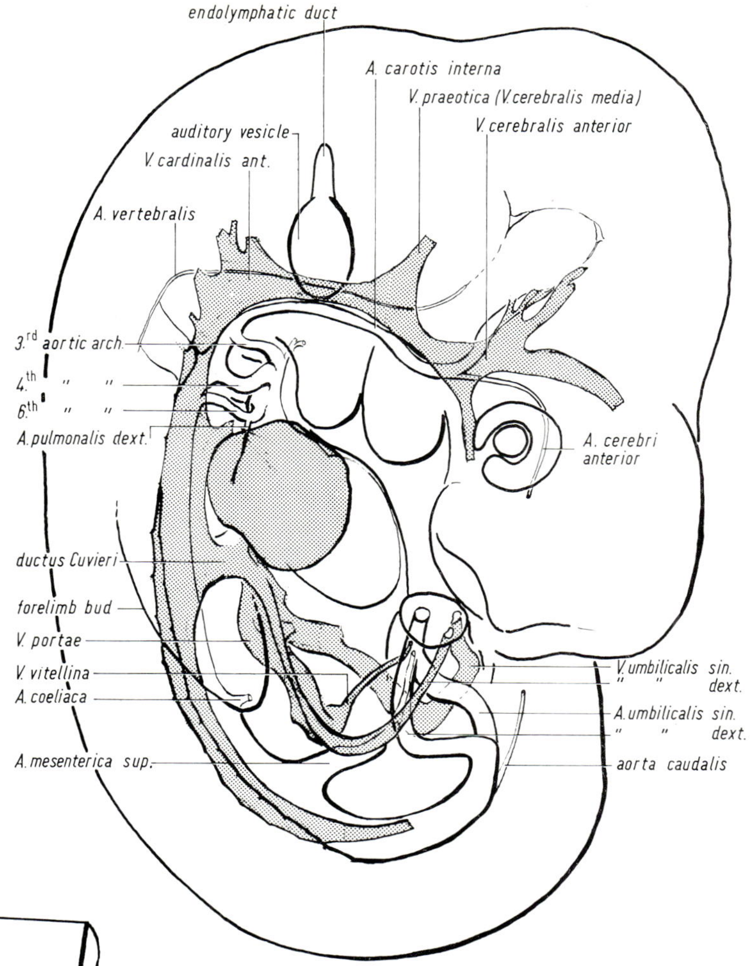

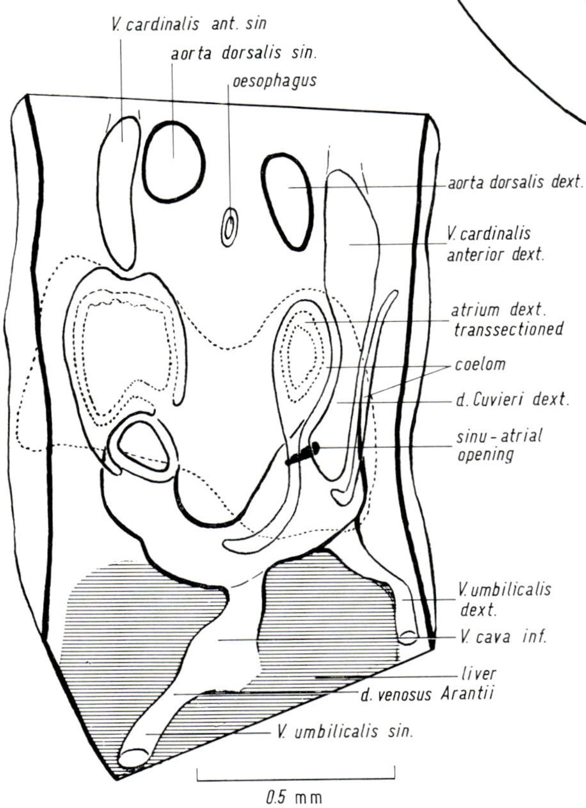

Fig. 173
Sinus venosus, dorsal view. Reconstruction.
Contours of atrium: stippled.
KT 630c/4

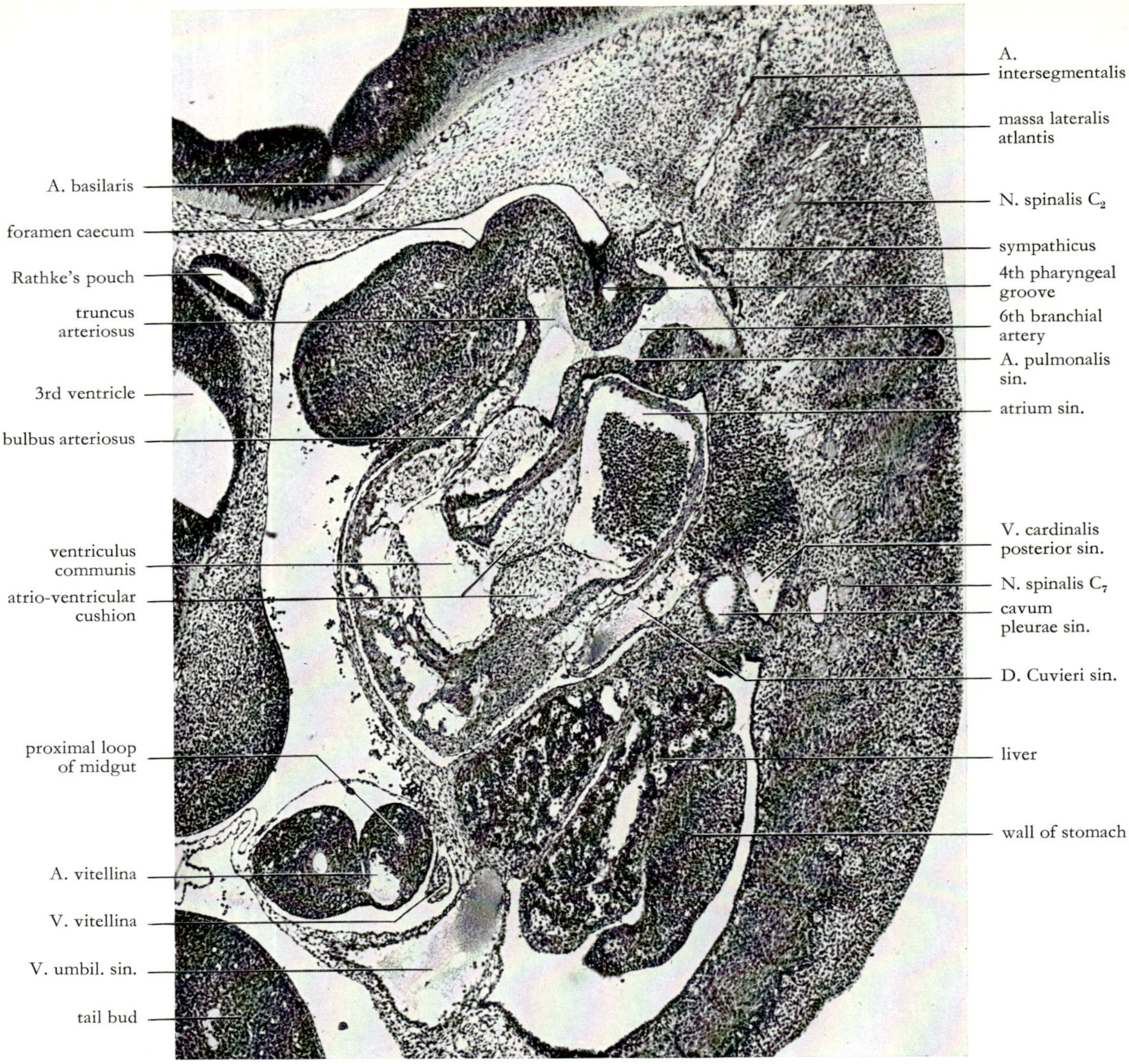

Fig. 175
Paramedian section. Nominal age 12 days, developmentally 11 $^1/_2$ days
KT 630b

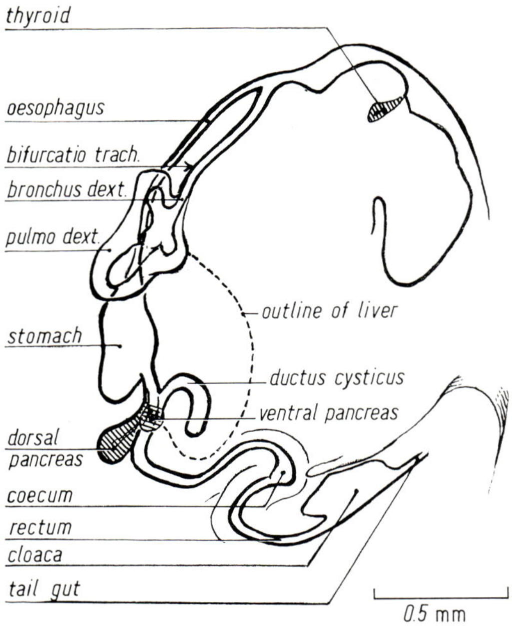

Fig. 176
Intestinal tract, reconstruction.
KT 630b, 6mm, developmental age 11 1/2 days

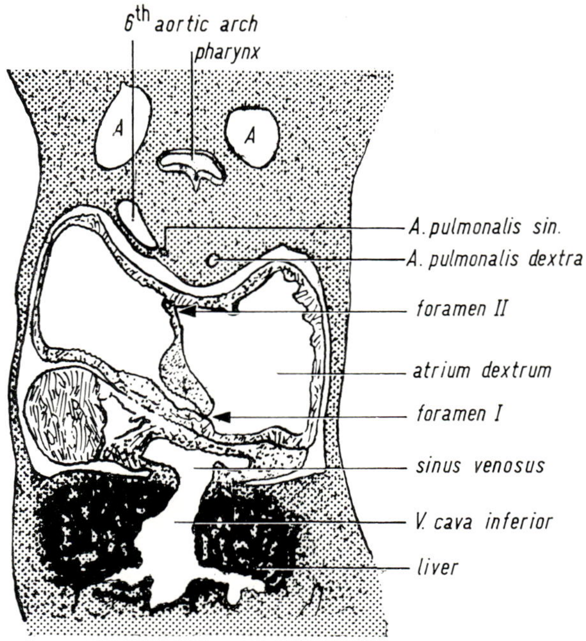

Fig. 174
Septation of atrium, frontal section, dorsal view.
A Aorta dorsalis (paired).
KT 630c/4

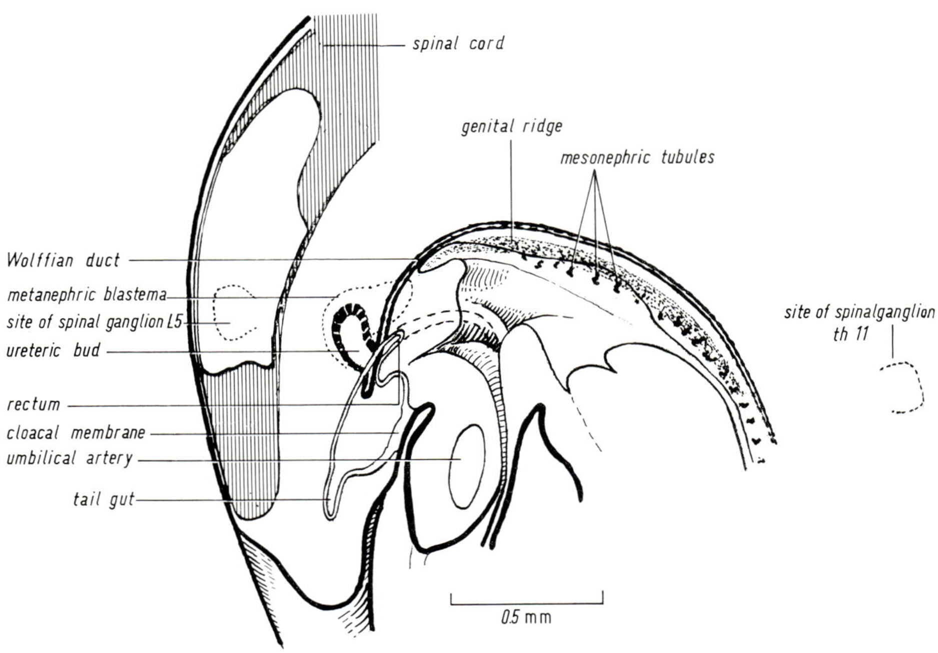

Fig. 177
Urogenital apparatus, left half, reconstructed.
KT 630b

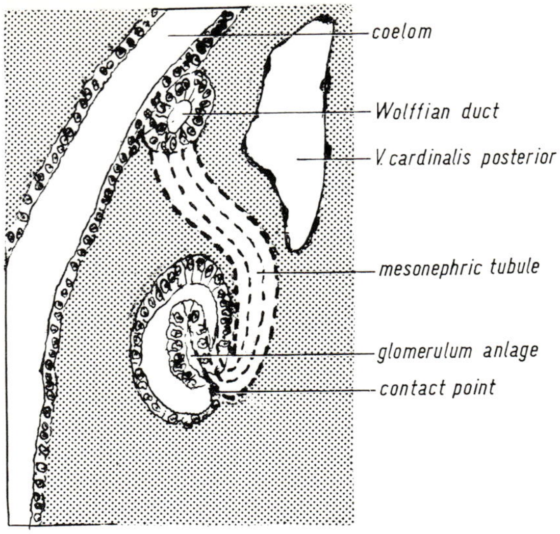

Fig. 178
Differentiation of mesonephros. Cross section.
KT 630a, developmental age 11 1/2 days

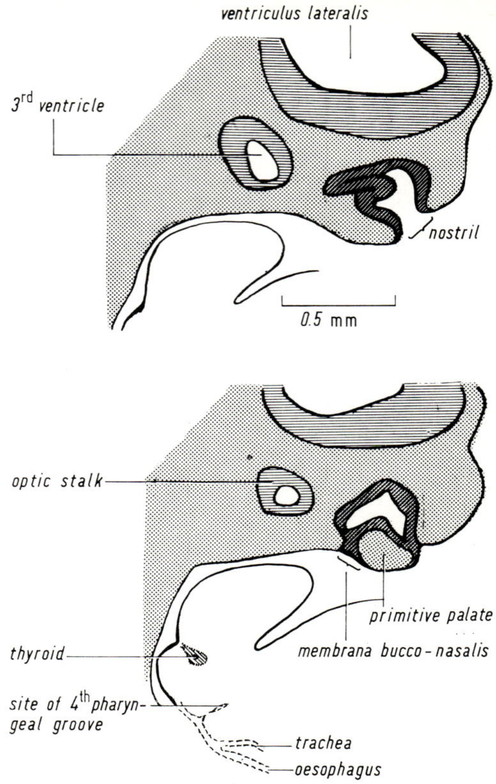

▲
Fig. 179
Development of olfactory apparatus.
Lateral section (above) and medial section (below), in sagittal direction.
KT 630b, developmental age 11 1/2 days

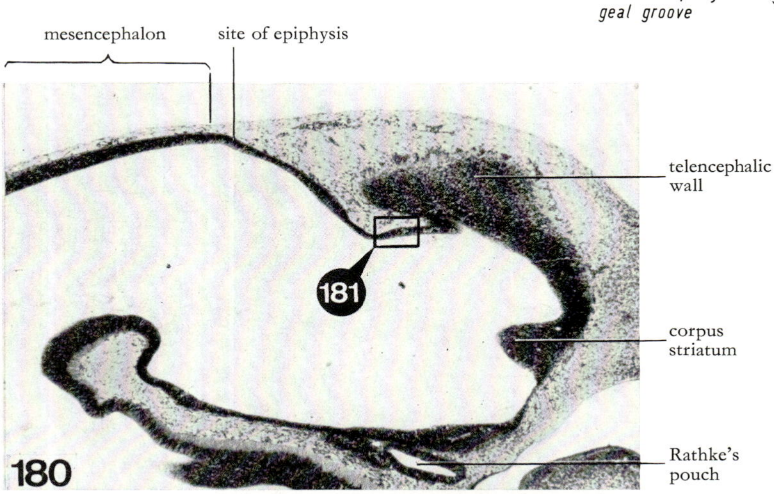

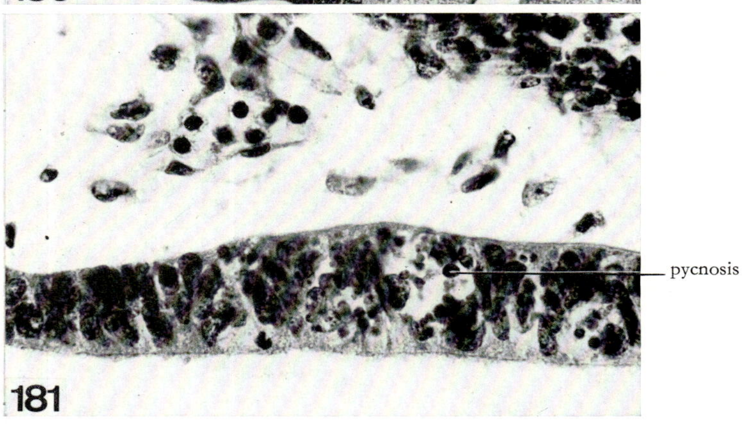

◄
Fig. 180
Sagittal section through fore- and midbrain.
KT 1005/1, 11 1/2 days. 40:1

Fig. 181
Detail of Fig. 180.
Desintegrating cells in the anterior roof of the diencephalon. 560:1.

Horizon XVII
homo = 41 days
11–13 mm

Stage 20 Earliest Sign of Fingers

12 Days, 7–9 mm

External Form

The most conspicuous changes take place in the extremities. The "handplate" is no longer roundish. Even in younger members of this group there is a slightly angular contour (Fig. 182). The developing angles correspond to the finger rays [184]. The posterior footplate is now demarcated from the lower part of the leg. The somites are clearly visible from the tail to the mid-trunk region. The spinal ganglia may be seen through the skin in transparent specimens.
Length: 7–9 mm, measured in a direct crown-rump line.

Circulatory System

Arteries: The second aortic arch has disappeared. The 3rd, 4th and 6th are of variable caliber, but symmetric (Fig. 191).
Heart: The truncus arteriosus is being partitioned. The septum membranaceum is still incomplete.
The septation of the atrium (Fig. 192) began in the preceding period. This important event is illustrated in Figs. 193–196. These sections were taken from embryo KT 630. Its nominal age is 12 days, but its developmental age is 11 1/2 days.

Intestinal Tract

Profound transformations in the intestinal tract are occurring.
In the *pharyngeal region*, the anlage of the *tongue* is delimited from the lower jaw by a furrow (Fig. 190).
The anlagen of the *teeth* appear as slight epithelial thickenings. A continuous dental lamina is not apparent anteriorly [76].
Lateral to the mid-sagittal plane, the furrow in front of the tongue gives rise to the outbudding ductus submaxillaris.
The *thyroid primordium* is closely attached to the arcus aortae. It is displaced very deep and its connection with the slight indentation of the foramen caecum is becoming indistinct (Fig. 197, dashed line). The cells of the thyroid are strongly eosinophilic. Sometimes, they seem to form rosettes. True formation of follicles, however, does not occur until a much later phase [147].
The 3rd and 4th *pharyngeal pouches* are now budding out. The 3rd produces the thymus and parathyroid anlage [146]. They still connect with the pharyngeal epithelium, but the uniting epithelial bridges contain many decaying cells.
The 4th pouch yields the so-called ultimo-branchial body [146].

The *lung buds* (Figs. 188, 189) have not only secondary (= lobar), but tertiary (= segmental) bronchi. The pleural cavity is in broad communication with the peritoneal cavity (Fig. 198, ductus pleuro-peritonealis).

The stomach is greatly distended, and there are regional differences in the epithelium (Fig. 188). Both rudiments of the pancreas are in contact with each other.

Within the *liver*, there are megakaryocytes, and as previously mentioned, blood is being formed.

Urogenital Tract

The *mesonephros* contains distinct tubules, but no well-formed glomeruli. Evidently mesonephric glomeruli remain primitive in mice, and perhaps they never function.

The *metanephros* has only two polar tubes (pelvic poles) in younger specimens (KT 643/2), but many secondary buds in older (KT 634/8) specimens of this group [102].

The *ureter* is relatively narrow and opens into the Wolffian duct which still ends blindly within the epithelium of the urogenital sinus.

The uro-rectal septum has not yet reached the cloacal membrane (Fig. 198).

Fig. 182 Embryo, from the right. Albino of a control series, fixed, on millimeter scale.
Tr Nasolacrimal groove. 10:1

Fig. 183 Embryo from the left, life photograph.
KT 942, 11 days 23 h. 9:1

Fig. 184 Explanation of Fig. 183
O Otic vesicle, *Ns* nostril, *F* choroid fissure, *I.* first branchial pouch, *Hl* hindlimb bud, *Tr* nasolacrimal groove

Fig. 185 Eye, frontal section.
Pi Pigment epithelium, *Lf* lens fibers.
KT 943/11, 12 days. 130:1

Fig. 186 Primordium of vertebral column, thoracic region. Frontal section.
N Spinal nerve, *Iv* intersegmental vessels, *If* intrasegmental fissure (sclerotomic fissure), *a* cranial sclerotomic half, *b* caudal sclerotomic half, *CH* notochord.
KT 941/3, 12 days, 8 mm length. 100:1

Fig. 187 Sagittal section through cervical ganglia.
N XII Nervus hypoglossus, Gc_2 ganglion cervicale 2, *At* atlas, *A.v.* arteria vertebralis.
35:1

Fig. 188 Sagittal section through lower thoracic and abdominal region.
Oe Oesophagus, *St* stomach, *Li* liver, *Pa* pancreas, *K* kidney, *Sy* sympathetic trunk, Th_5 caput costae 5, *V* vena cardinalis posterior sinistra.
KT 943/8, 12 days. 40:1

Fig. 189 Detail of Fig. 188.
Lu left lung, *Sp* nervus splanchnicus, *NN* suprarenal gland. 100:1

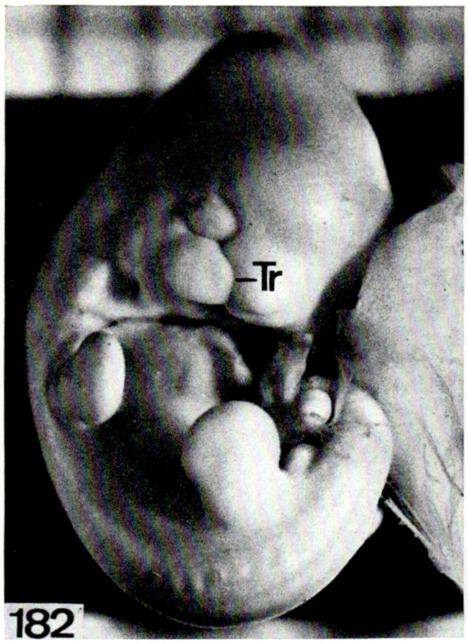

182

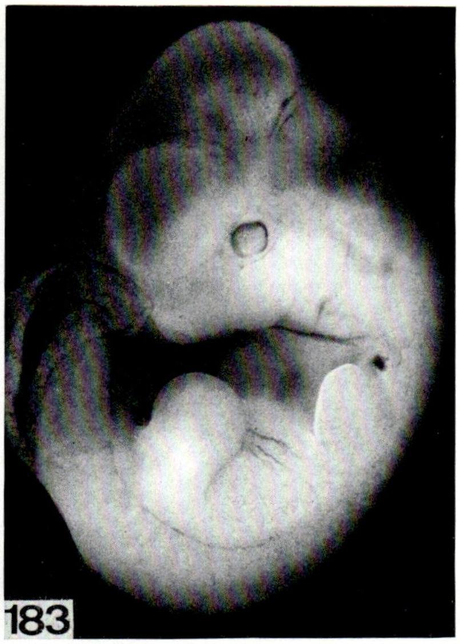
183

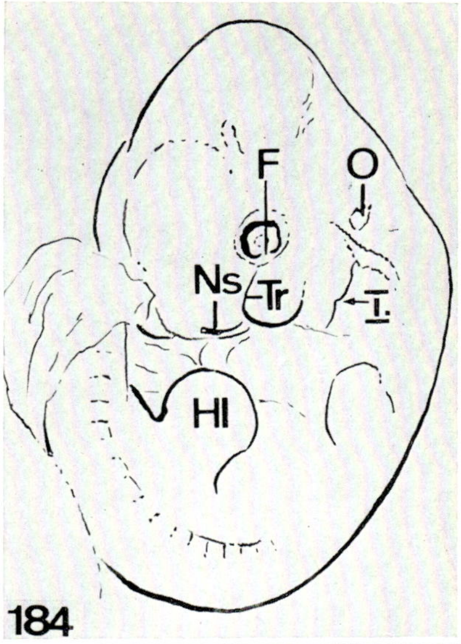

184

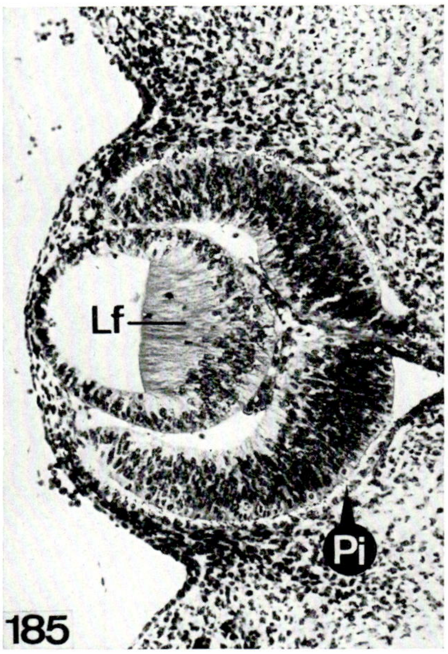

185

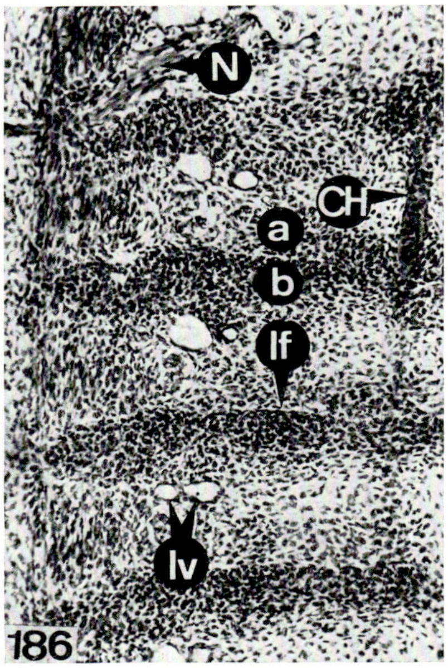

186

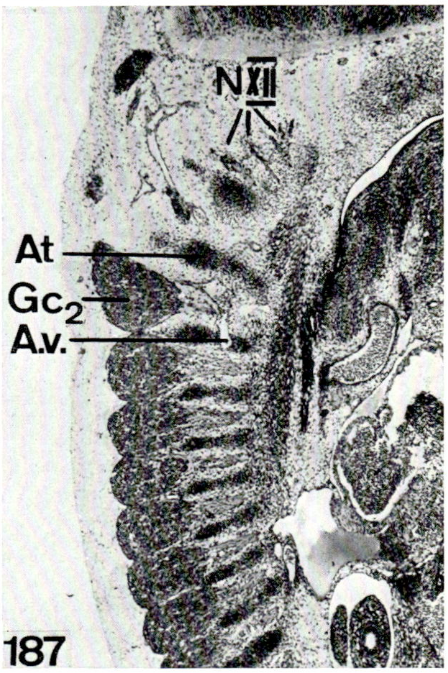

187

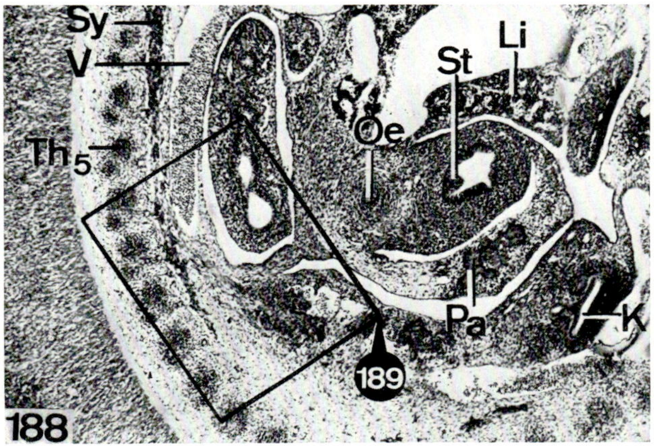

188

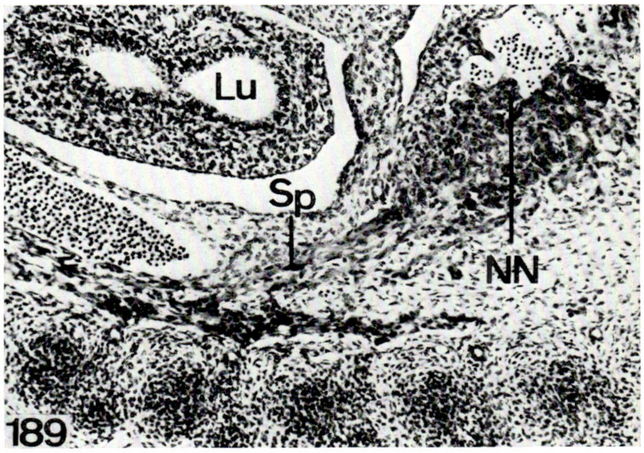

189

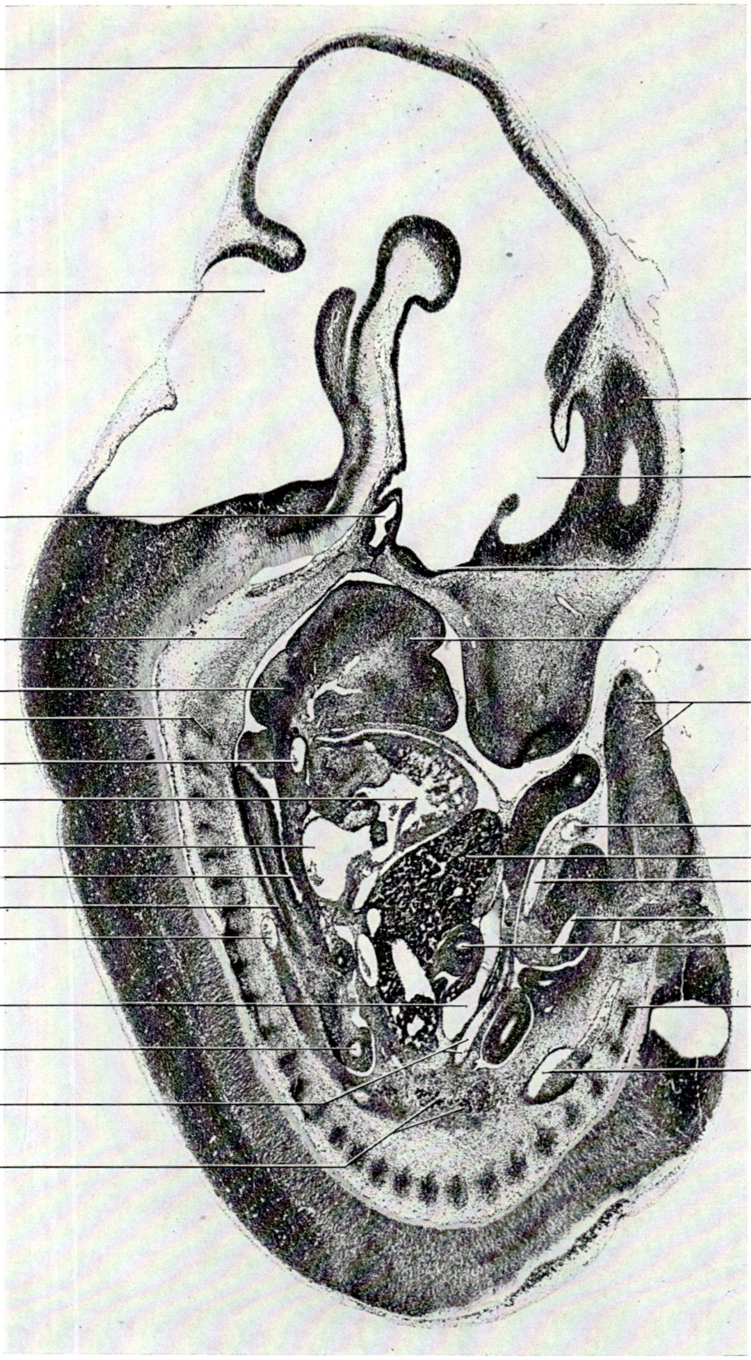

Fig. 190
Sagittal section.
KT 943/8, 12 days, 8 mm length

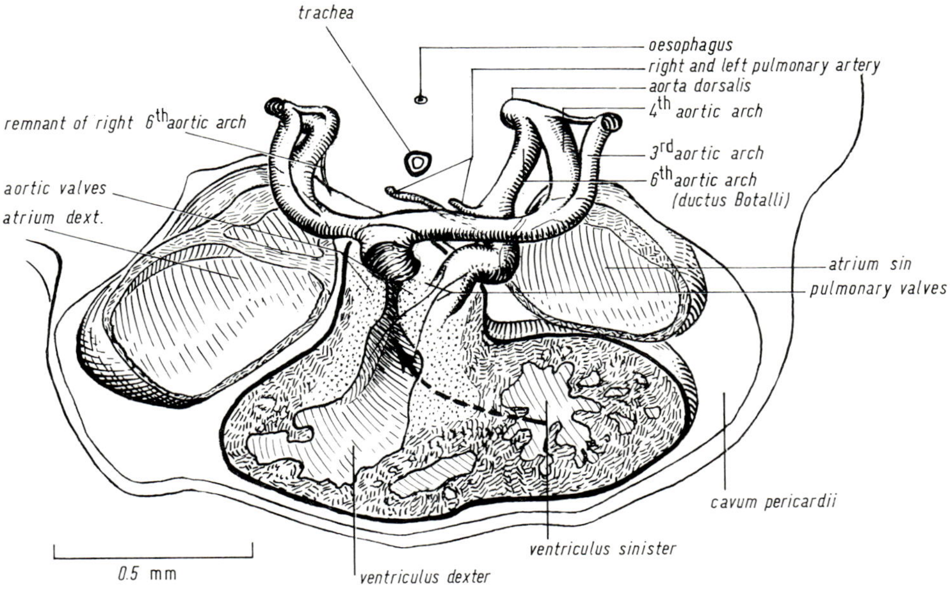

Fig. 191
Frontal section and reconstruction of the arterial trunk.
Arrow: in the aortal pathway. Septum membranaceum still incomplete. Beginning formation of semilunar valves.
KT 943/11, 12 days

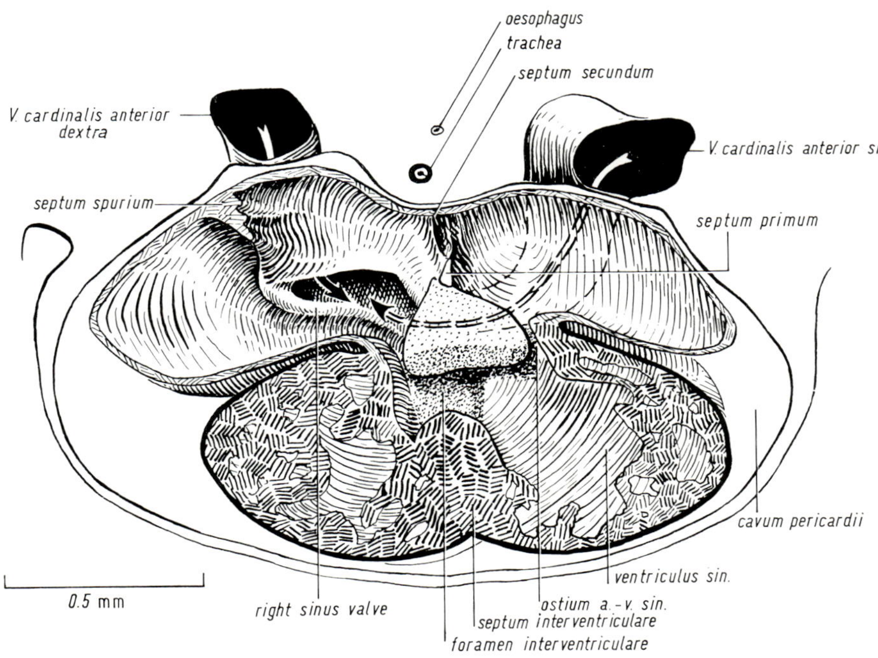

Fig. 192
Frontal section and reconstruction of the atrium.
Arrows: indicating the blood flow of the sinus venosus. Upper atrio-ventricular cushion = clear, lower cushion = dark (stippled).
KT 943/11, 12 days

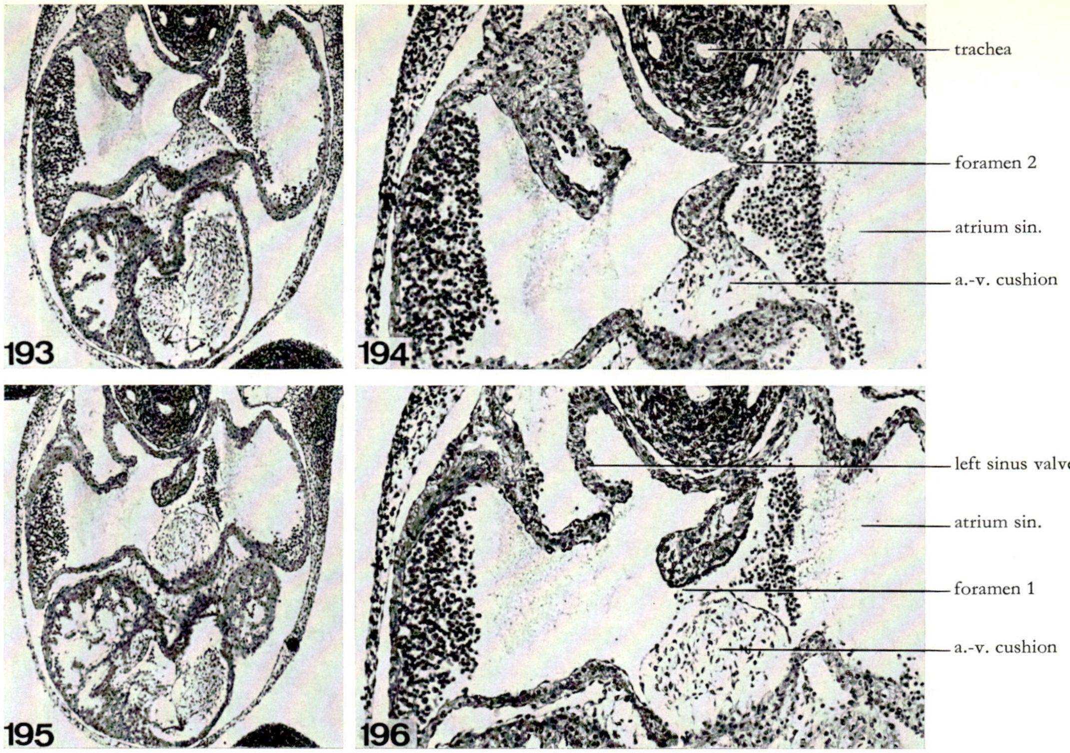

Figs. 193–196
Low and medium magnification of cross sections through the heart. Developmental age 12 days. Formation of foramen primum and secundum.
KT 630, 55:1 and 100:1

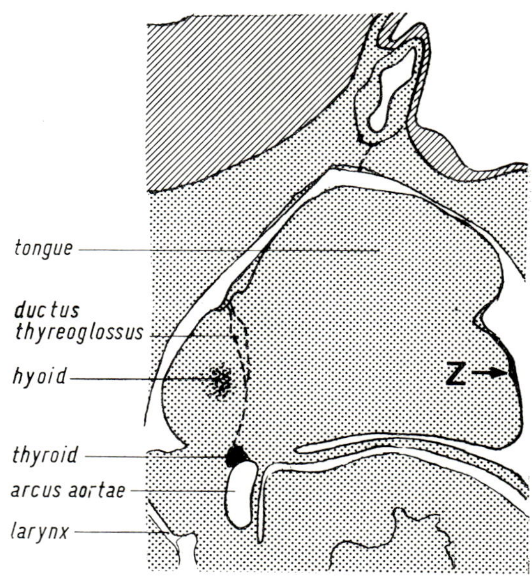

Fig. 197
Sagittal section of oral and pharyngeal region with thyroid primordium.
The thyreo-glossal duct (broken line) is interrupted, *Z* thickening of the mandibular epithelium. Lip furrow and lower incisivus will develop from here

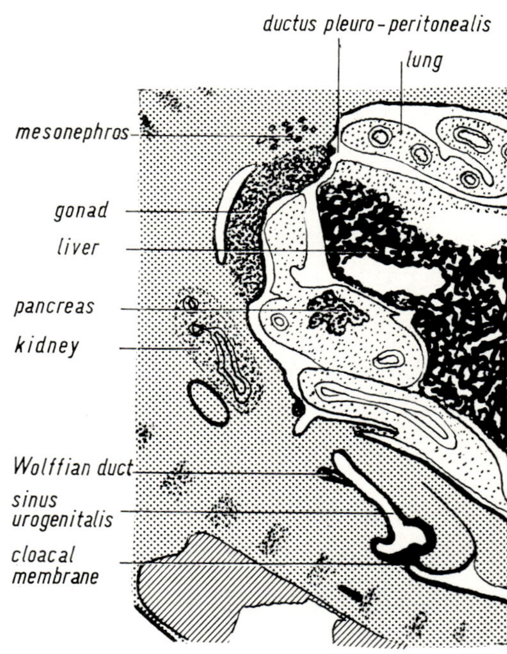

Fig. 198
Urogenital apparatus, sagittal section.
KT 943/8, 12 days, 8 mm length

Fig. 199
Reconstruction of the labyrinth, sagittal section. The major part of the lateral and upper semicircular ducts are cut away. KT 943/7, 12 days, 8 mm length

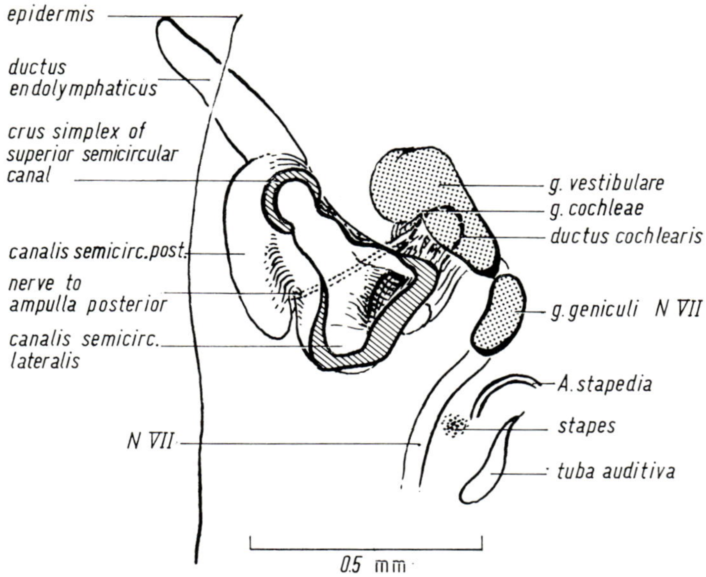

The *gonads* are still in the indifferent state (Fig. 198). Sexual differentiation is usually apparent histologically at 12 1/2 days and sometimes at 12 days [118].
The *suprarenal* is distinct and is composed of cellular cords representing the cortex [137] (Fig. 189).

Central Nervous System

In this phase, the pineal gland appears as a discrete evagination in the most posterior part of the diencephalic roof (not visible in the section shown in Fig. 190). In the anterior part of the roof, near the interventricular foramen, there are still numerous pycnoses [166].
Eye: The posterior wall of the lens vesicle is markedly thickened as a result of differentiation of lens fibers (Fig. 185).
Otocyst: The subdivisions of the labyrinth are easily recognized. The semicircular canals, however, are still flat pouches, which are not yet tubular. The cochlear duct is short (Fig. 199, dashed line). The vestibular ganglion has larger cells than the cochlear ganglion (most of which is hidden in Fig. 199).
The auditory capsules still consist of mesenchyme, which will soon chondrify.

Material	Age	
KT 620–21	11 days 22 h	6 embryos: 8, 8, 8.5, 8.5, 8.8 mm
KT 941–43	11 days 23 h	5 embryos: 7.6–8.3 mm
KT 622–23	12 days	7 embryos: 7, 7.5, 8.2, 8.6, 9, 9, 9.1 mm
KT 628	12 days 3 h	5 embryos: 6–8 mm
KT 629	12 days 3 h	6 embryos: about 8 mm
KT 636	12 days 2 h	7 embryos: 7.5–8.7 mm

Stage 21 Anterior Footplate Indented; Marked Pinna

Horizon XVIII–XIX
homo = 44–48 days
14–20 mm

13 Days, 9–11 mm

Externally: The rapid development of the *pinna* is striking, and it now forms a crest at right angles to the head. Five rows of whiskers are present. The 4 upper rows of whiskers are shown in Fig. 202, as well as a prominent hair follicle above the eye and another in front of the ear. The indentation of the "*handplate*" is characteristic of this period. It is shallow in younger embryos of this stage (Fig. 200) and very distinct in older ones (Fig. 201). The footplate is just becoming indented.
Somites are clearly visible only in the distal part of the tail. Usually the tail is curved to the left (as in Fig. 202) rather than to the right.
Length: The length varies from 9–11 mm in unfixed embryos.
Sagittal section (Fig. 208): In the brain, the choroid plexus projects into the lateral and the 4th ventricle, forming finger-like evaginations. The tongue projects from the floor of the mouth [76].

Locomotive System

The extremities now differentiate rapidly [184]. As shown in Fig. 209, the skeleton of the forelimb already contains cartilage, while the "hand" still remains mesenchymous. The cranial vertebrae are chondrified, but not the more posterior ones.

Circulatory System

The aortic and pulmonary trunks are completely separated as shown in Fig. 208. The membranous part of the interventricular septum is not yet closed. Within the coronary sulcus, fine coronary vessels are visible. There is probably not yet a continuous circulation. All valves of the heart are now present in primitive form.
The inferior caval vein is represented proximally by a large hepato-cardiac channel. The venous system is essentially the same, as that shown in the reconstruction (Fig. 224) made for the next developmental phase (14 days).

Intestinal Tract

The *oral cavity* is in broad communication with the nasal cavity. The palatine processes consist of mesenchyme and are in a vertical position. The dental lamina of the future molars is clearly visible [73] (Fig. 210).
In sagittal sections (Fig. 208) the *epiglottis* is delimited by a discrete cleft from the rest of the larynx which projects plug-like into the pharynx.
The *lungs* are clearly subdivided into lobes [61], and the segmental bronchi are continuing to branch.

The *thymus* has completely lost its connection with the pharynx. Numerous blood vessels are penetrating the thymus primordium, but the closely neighboring and considerably smaller *thyroid* is still a solid complex [147].

Within the abdomen, the *liver* is well developed and contains scattered blood-forming foci. The stomach does not yet have differentiated glands. In some places the epithelium is higher than in others, and there are some small lumina (Fig. 208). The spleen appears in cross sections as a triangular structure (Fig. 206).

The pancreas produces numerous sprouts which considerably swell the dorsal mesentery (Fig. 208).

Urogenital Tract

The urogenital tract is characterized by the rapid development of the *kidney* and by *sexual differentiation.*

The mesonephros contains many regressing mesonephric tubules (Fig. 204). The slender ureter is continuous with the distended pelvis, which has well-marked primary calyces with caps of metanephric tissue (Fig. 206).The metanephric caps [100], in some places, form small vesicles, which are more intensely eosinophilic than the ureteric bud. The vesicles are sometimes transformed into S-shaped tubules, the free ends of which have differentiating glomeruli. The *Wolffian duct* is more developed in the male, the *Müllerian duct* more in the female. The cloaca is completely subdivided. The ureter has contact both with the Wolffian duct and with the urogenital sinus. It still ends blindly, slightly cranial to the lower end of the mesonephric duct.

The sex of the gonads can now be diagnosed. In the male, the future seminiferous tubules appear as solid, regularly arranged strands [118]. They are composed of small supporting and nourishing cells, and of large primordial germ cells (Fig. 207). The first suggestion of sexual differentiation appeared at 12 1/2 days. Male primordial germ cells were located centrally, and female germ cells were located peripherally. Some of the large female gonocytes (Fig. 205) are dividing.

Central Nervous System

There are pronounced changes in the diencephalic roof (Fig. 211). There are no longer pycnotic cells in the anterior portion of the roof [166]. The choroid plexus is fully developed. The *pineal gland* was recognizable at 12 days as a small evagination, and it is now very distinct. The posterior commissure is also well formed [160].

The *hypophysis* is developing very rapidly (Figs. 212–214).

The structure of the *eye* is very characteristic for this group: The lens vesicle has lost its lumen, and is a solid sphere. The vitreous body is exceedingly small. In the retina, the layer of nerve fibers forms a wide border with peripherally tapering ends. The nuclear layers of the retina are still predominantly indistinguishable from each other. However, the clear nuclei of the future ganglionic cells are discernible in some places. The ganglion cell layer is the first one to be delimited from the other layers.

Material	Age	
KT 615–16	12 days 23 h	7 embryos, 10–11 mm
KT 633–34	13 days 3 h	6 embryos, 10–10.8 mm
KT 1014–16	13 days 2 h	8 embryos + 1 resorption, about 10 mm
KT 901	12 days 20 h	7 embryos, 9.6–10.0 mm (after fixation)

Fig. 200 Embryo from the right. Younger stage of 12 days 6 h. Formalin fixed.
Ey Eye, *Tel* cerebral hemisphere.
KT 952, 8.7 mm length. 8:1

Fig. 201 Embryo from the left, 13 days, 10 mm length, life photograph.
Si Sinus sigmoideus.
KT 634. 7:1

Fig. 202 Frontal view. Albino of control series. Bouin fixed, 13 days, 10 mm length.
O Pinna, *So* tail somites, *N* nostril, *H* rudiments of hair follicles (whiskers). 10:1

Fig. 203 Eye, cross section, 13 days.
OL Upper lid, *L* lens, *St* optic stalk, *P* pigment layer. 70:1

Fig. 204 Ovary and vicinity. Cross section, 13 days.
NN Suprarenal gland, *U* mesonephric tubules, *M* Müllerian duct, *Ma* stomach.
KT 901/5. 100:1

Fig. 205 Detail of Fig. 204.
Oz Oocytes in meiotic prophase. 550:1

Fig. 206 Testis and vicinity, 13 days.
NN Suprarenal gland, *Ur* ureter in kidney rudiment, *W* Wolffian duct, *M* Müllerian duct, *B* omental bursa, bordering the triangular anlage of the spleen (right) and the stomach (below).
KT 901/2

Fig. 207 Detail of Fig. 206, testis.
Sp Gonocytes. 550:1

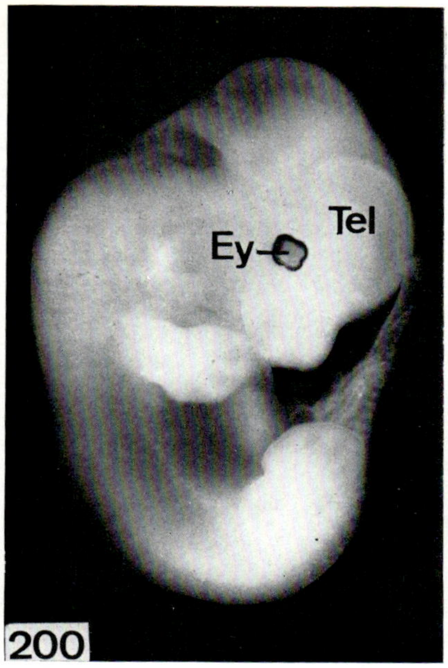
Ey
Tel
200

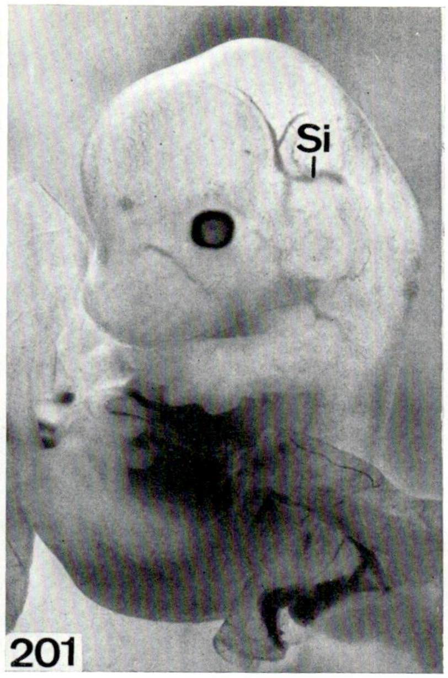
Si
201

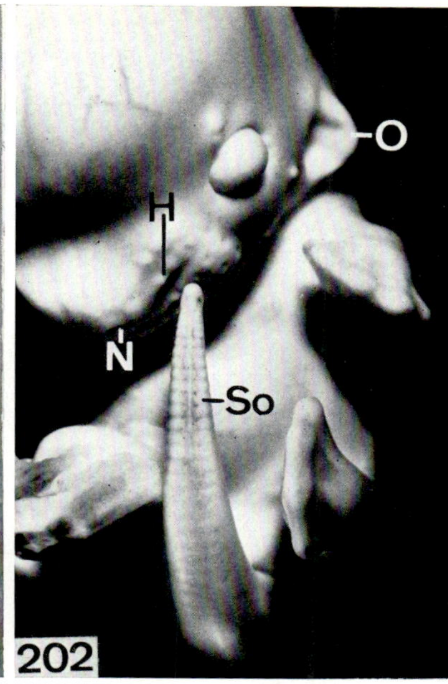
O
H
N
So
202

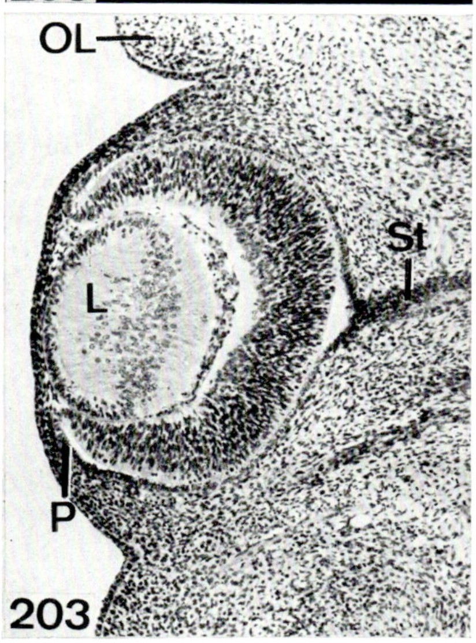
OL
St
L
P
203

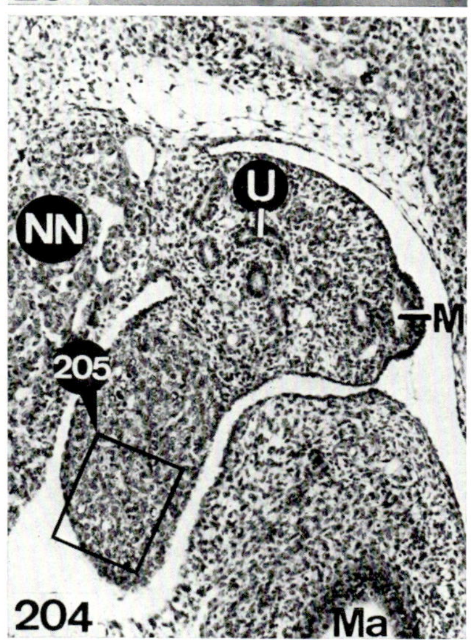
U
NN
M
205
Ma
204

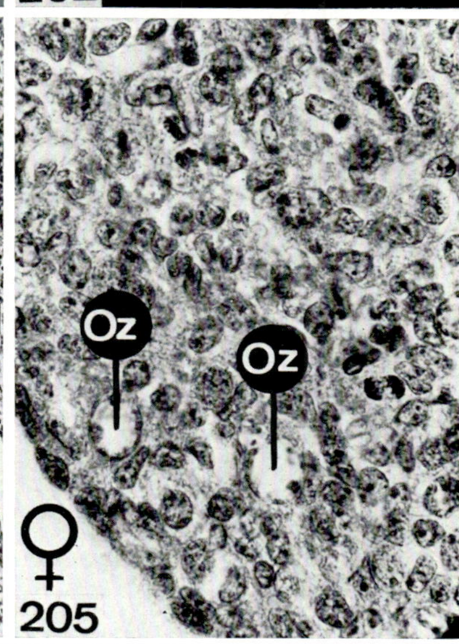
Oz
Oz
♀
205

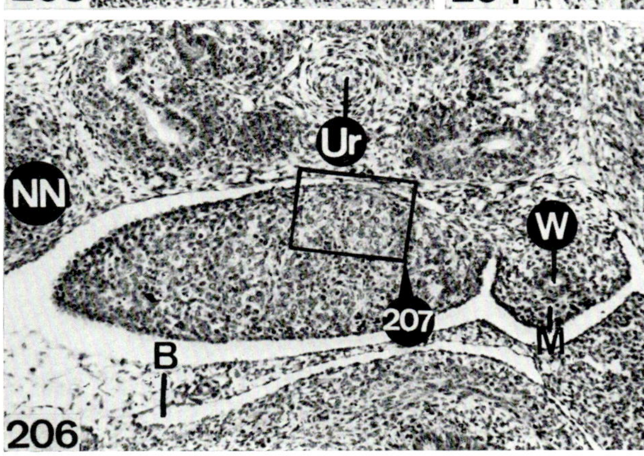
Ur
NN
W
207
M
B
206

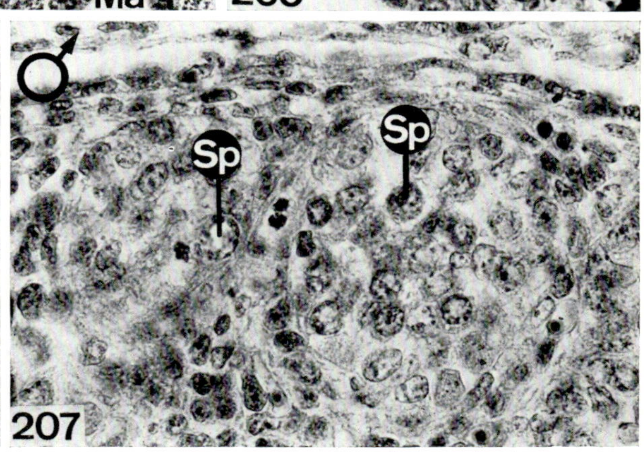
♂
Sp
Sp
207

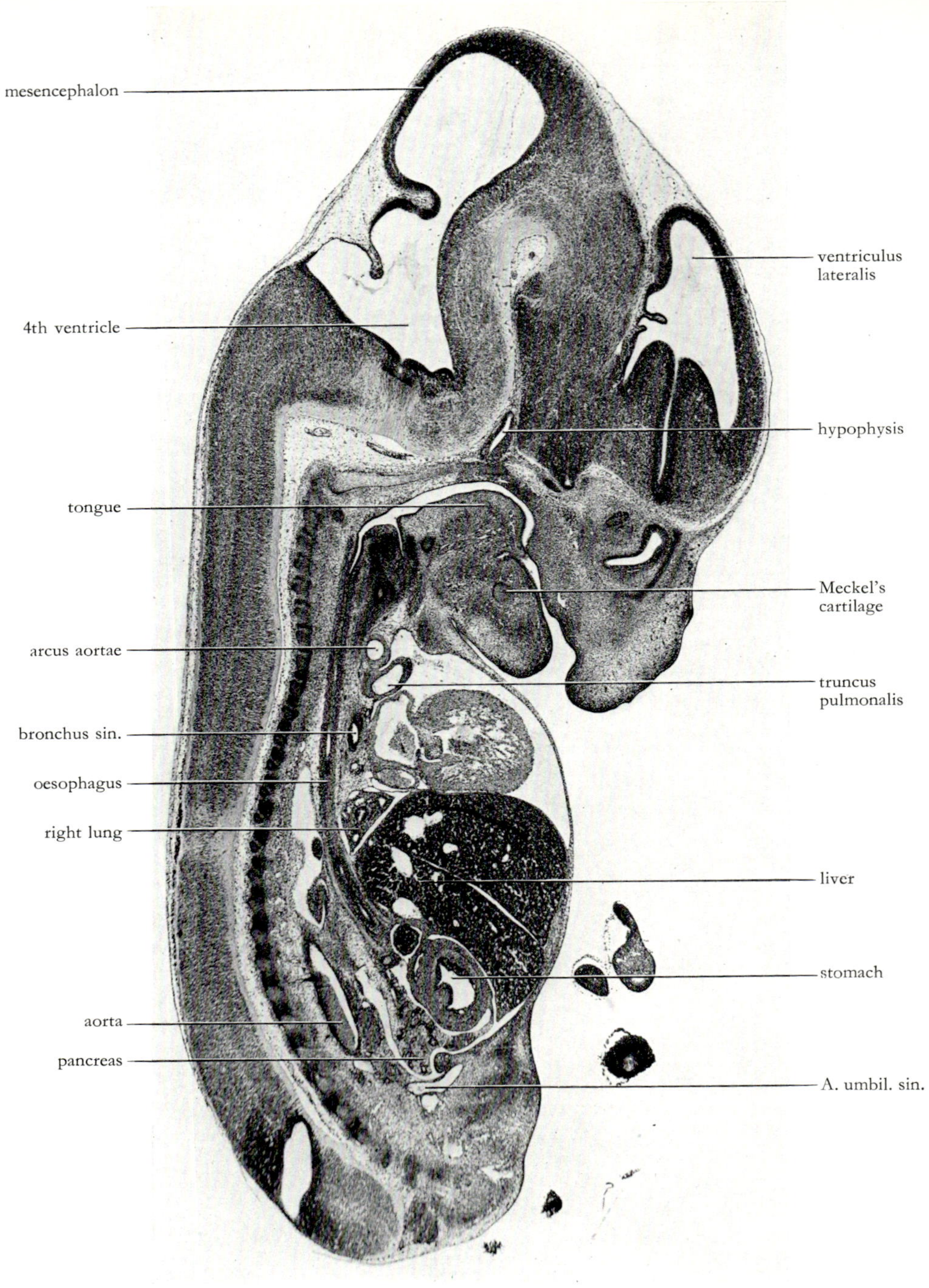

Fig. 208
Sagittal section, 13 days, 10 mm length.
KT 901/2

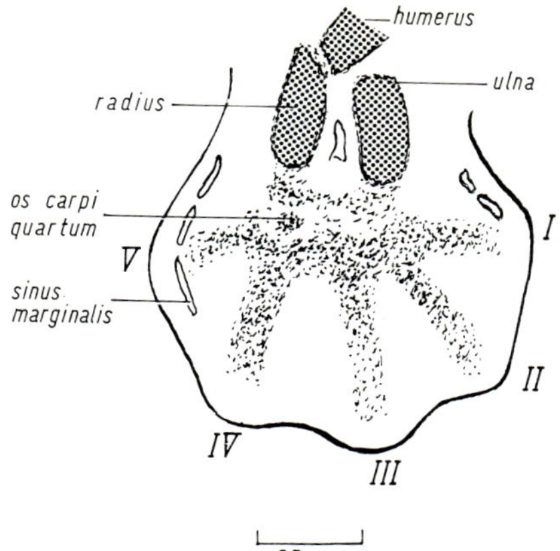

Fig. 209

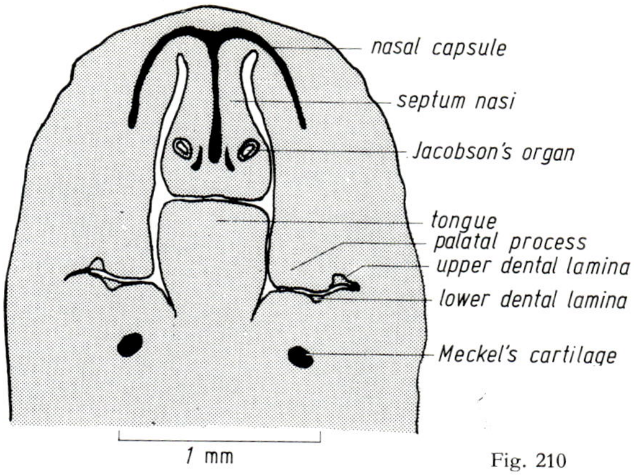

Fig. 210

Fig. 209
Right forefoot plate, 13 days.
Precartilage (of "arm" skeleton): darkstippled. Mesenchymal condensations: light-stippled.
KT 1014

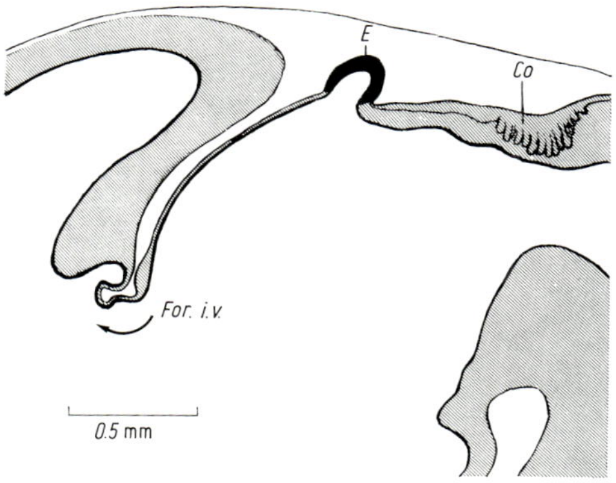

Fig. 210
Cross section through oral and nasal cavities. Plane of section nearly horizontal, therefore the conchae are not visible.
Cartilage and precartilage: black.
KT 901, 12 days 2 h

Fig. 211
Sagittal section through roof of diencephalon.
E Epiphysis, *Co* commissura posterior, *For i.v.* foramen interventriculare.
KT 633/b, 13 days

12 days | 12 1/2 days | 13 days

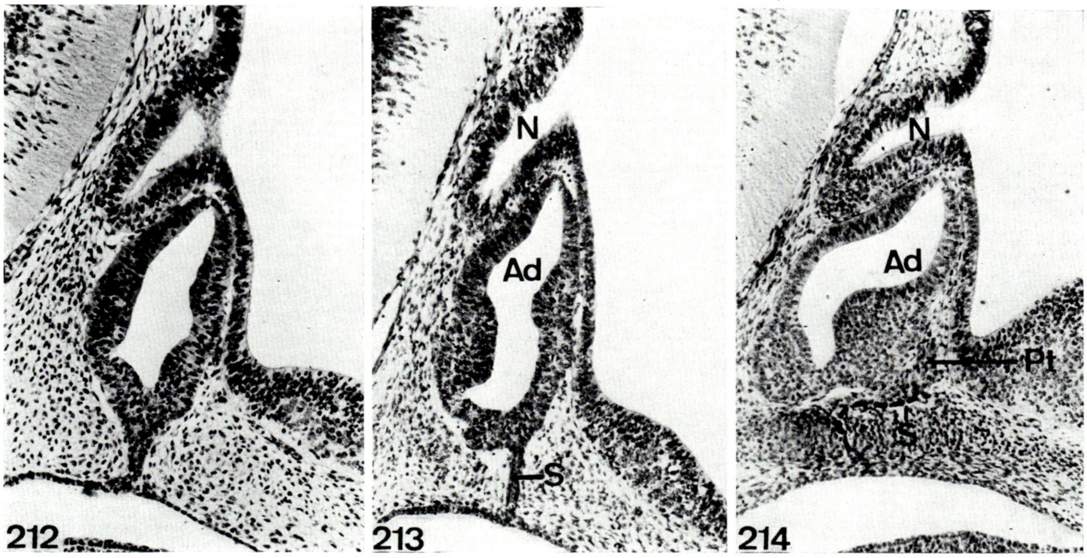

Figs. 212–214
Development of hypophysis. Sagittal sections, 12–13 days.
N Neurohypophysis, *Ad* adenohypophysis, *P.t.* pars tuberalis, *S* connecting stalk (with pharyngeal roof)

Horizon XX–XXIII
homo = 51–57 days
20–32 mm

Stage 22 Fingers Separate Distally

14 Days, 11–12 mm

Externally: With magnification, one can easily see that the individual *fingers* are separated in the fore-foot plate. In contrast, in the hind-foot plate there are deep indentations between the developing toes, but they are not yet separated (Figs. 215–217). Numerous young *hair follicles* may be recognized in the skin, except for the head region. The *somites* are discernible only in the distal part of the tail. The growing pinna is turned forward and covers about one half of the external auditory meatus. The umbilical hernia is very conspicuous at this stage (Figs. 216 and 217).
Length: The length varies from 10.4 to 12 mm.

Circulatory System

Arteries: Both *umbilical arteries* are united at the umbilical ring to form a single vessel which passes through the umbilical cord (not visible in Fig. 224).
The arterial system of the *head* is being transformed. In Fig. 224, the *arteria stapedia* is recognizable. It is a branch of the 2nd aortic arch. The stump of the right pulmonary artery is visible below the ductus arteriosus Botalli. The right 4th aortic arch has lost its communication with the dorsal aorta, and flows directly into the artery of the right forelimb (arteria subclavia dextra). On the whole, the definitive pattern of the prenatal circulatory system is established, and, as well, the definitive shape of the heart. The ventricular septum is now closed.
Veins: The *cardinal veins* are markedly asymmetrical. The posterior cardinal vein as shown in Fig. 224 is the original *left* posterior cardinal vein, while the right has regressed (in the reconstruction of a 12-day stage–Fig 172–both are present). The anterior cardinals are both present and will persist in the adult mouse.

Intestinal Tract

The palatal processes are elevating and begin to separate the oral and nasal cavities (Figs. 225 and 226). They are not yet fused, so that the cavities are still continuous. Occasionally, the fusion of the palatal processes may be delayed and the tongue appears to be squeezed between them (Fig. 223). The posterior part of the open communication in Fig. 225 always remains open. With progressive growth, this communication will be displaced posteriorly and form a long duct (ductus naso-pharyngeus).
The bud of the first *molar* [73] is visible (Fig. 226). The second and third will appear later. In sagittal section (Fig. 223), the anlagen of the lower and upper incisors [76] may be recognized (not labeled). They are located above and beneath the tip of the tongue.
Meckel's and Reichert's cartilage are more advanced in development than the precartilaginous skeleton of the *larynx*. The tracheal rings are still mesenchymal condensations.

The *thymus* and *thyroid* are easily recognizable (Fig. 223). The thymus has lost its connection with the 3rd pharyngeal pouch, and is situated above the pericardium. The parathyroids are joining the thyroid [146].
The bronchi are considerably distended and have numerous ramifications (Fig. 223).
The gut projects into the wide umbilical hernial sac. The mucosal lining of the gut consists of relatively tall columnar cells. The configuration of the intestinal tract is represented in Fig. 227.

Urogenital System

The *ureter* now opens into the urogenital sinus separately from the opening of the Wolffian duct. The sinus itself has been separated from the rectum in an earlier phase of development. The topography of the embryonic urogenital system is represented in Fig. 228, and microscopic details are shown in Fig. 223.
Sexual differentiation is very apparent: The seminiferous tubules are solid strands of cells which are nearly symmetrically arranged [118] (Fig. 221). The ovary, on the other hand, does not form cords [122]. In the female, the Müllerian duct is now more highly developed than the Wolffian duct (Fig. 219).

Central Nervous System

In the *telencephalon* cells migrate from the mantle zone and form a *superficial cortical layer* (primary cortex). It is separated from the broad mantle zone by a thin marginal zone, with few nuclei (Fig. 223). The choroid plexus of the lateral ventricle projects far into the lumen. All cranial nerves are now distinctly visible, and the ganglia are relatively large (Fig. 229).
Eye: The cavity of the vitreous body has increased considerably in size. The ganglionic layer of the retina (not the bipolar layer) appears as a zone of clear nuclei. It is distinguishable from the two darker inner zones which are not yet separated from each other. The separation occurs much later, after birth.
The eyelids are prominent.
Ear: The ductus cochlearis has elongated and it curves to form a full circle. It is bordered by a thin cartilaginous capsule. The presumptive sensory epithelium may be recognized by its height.
Hypophysis (Figs. 230–232): The connecting strand of epithelial cells (marked "*S*" in Fig. 214) has disappeared. The pars tuberalis continues to elongate. Ventral to the hypophysis is the cartilaginous base of the skull.
Placenta (Figs. 233–234): The labyrinth reached the peak of its development at 12 days.
The development of the so-called "yolk sac diverticles" is a conspicuous characteristic of this stage. They arise as clefts within the trophoblastic labyrinth and may perforate it [33] (Fig. 234).
The visceral (proximal) layer of the yolk sac has many branching folds.

Material	Age	
KT 618	13 days 21 h	4 embryos, 11–11.5 mm
KT 642–44	14 days	7 embryos, 11.5–12 mm (1 of them exencephalic)
KT 659–660	14 days	5 embryos, 10.5–12 mm
KT 1017–22	14 days 2 h	5 embryos, 10.6–11.9 mm
KT 1043–45	14 days 4 1/2 h	4 embryos, 10.4–11.3 mm

Fig. 215 Embryo from the left. Life photograph.
KT 642, 14 days, 12 mm length. 6:1

Fig. 216 Embryo from the left (albino of control series). Bouin fixed, 14 days, 11 mm length. 6.6:1

Fig. 217 Explanation of Fig. 216
O Pinna; *H* hair follicles; *Fi* finger, separated; *Nb* umbilical hernia; *So* tail somites

Fig. 218 Horizontal section through eye.
G Ganglion nervi optici, *K* nuclear layers (dark), *P* pigmented layer.
KT 643, 14 days. 55:1

Fig. 219 Ovary and vicinity, sagittal section.
W Wolffian duct, *M* Müllerian duct, *U* mesonephric tubules, *Ma* stomach, *DM* Mesogastrium dorsale.
KT 643b, 14 days. 100:1

Fig. 220 Detail of Fig. 219
Oz Oocyte, *Ep* epithelium, *Er* erythrocyte in capillary. 700:1

Fig. 221 Testis, tangential section. Seminiferous tubules, solid, nearly symmetrical arrangement.
Li Liver.
KT 643a, 14 days. 135:1

Fig. 222 Enlarged view of seminiferous tubule.
Sp Gonocyte, in prophase. 700:1

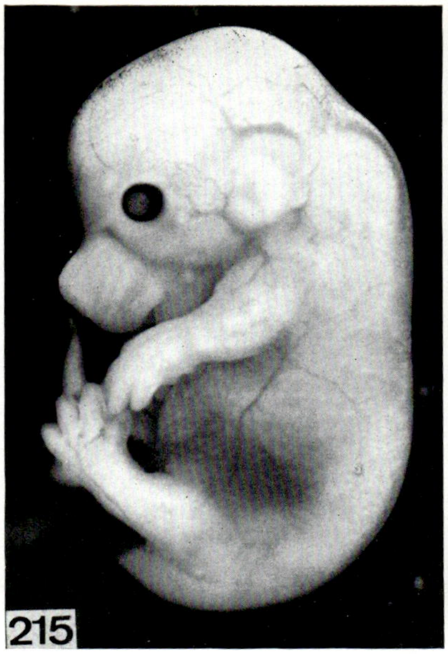
215

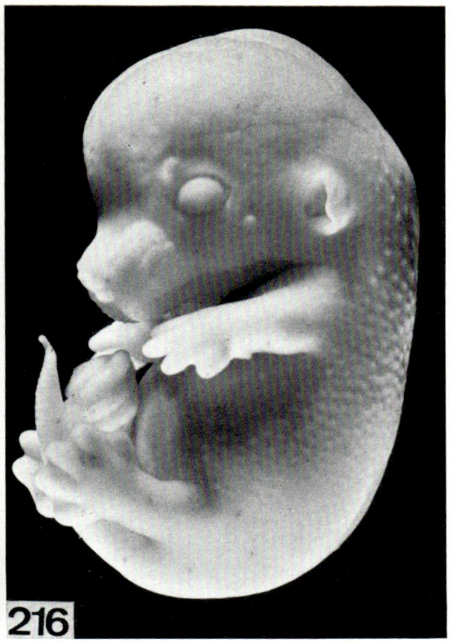
216

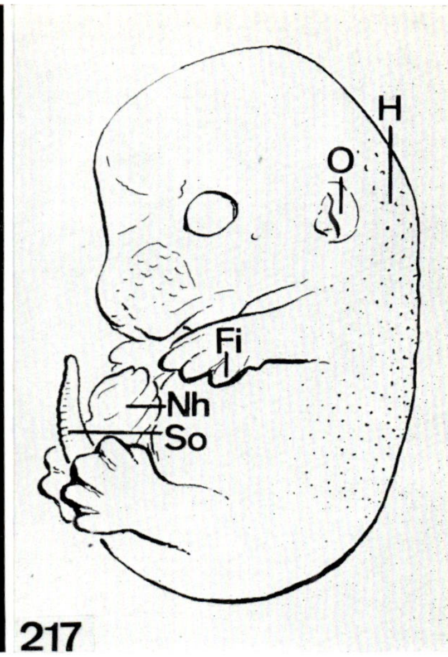
H
O
Fi
Nh
So
217

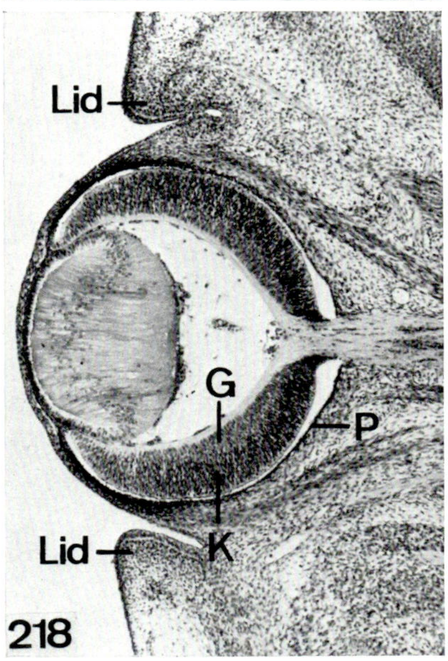
Lid
G
P
K
Lid
218

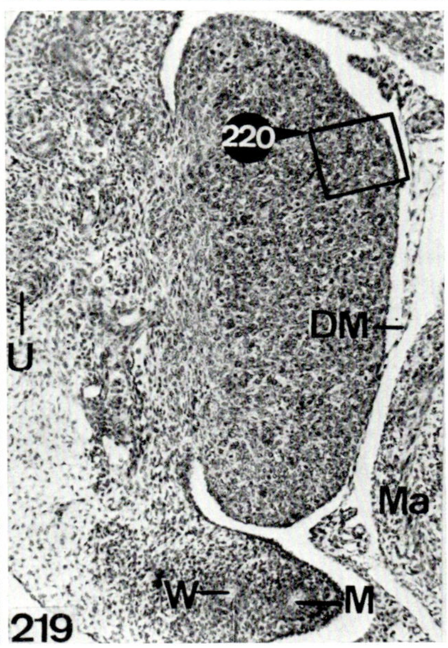
220
U
DM
Ma
W
M
219

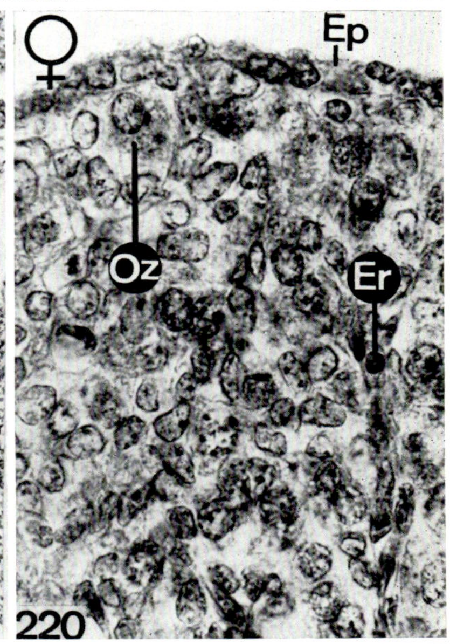
♀
Ep
Oz
Er
220

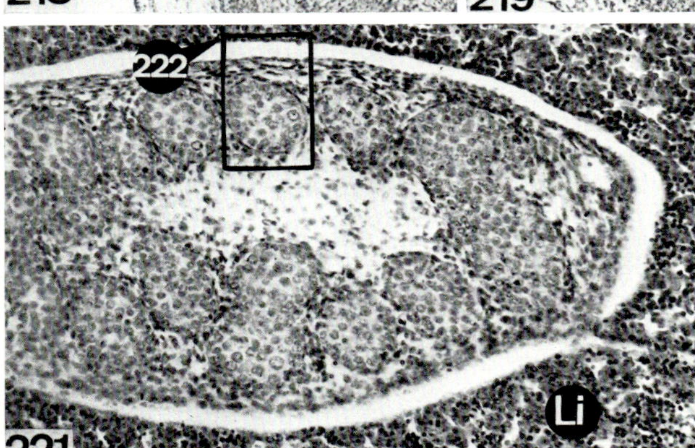
222
Li
221

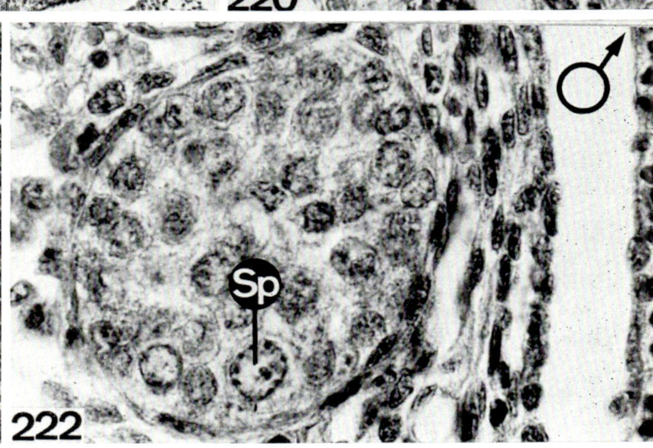
♂
Sp
222

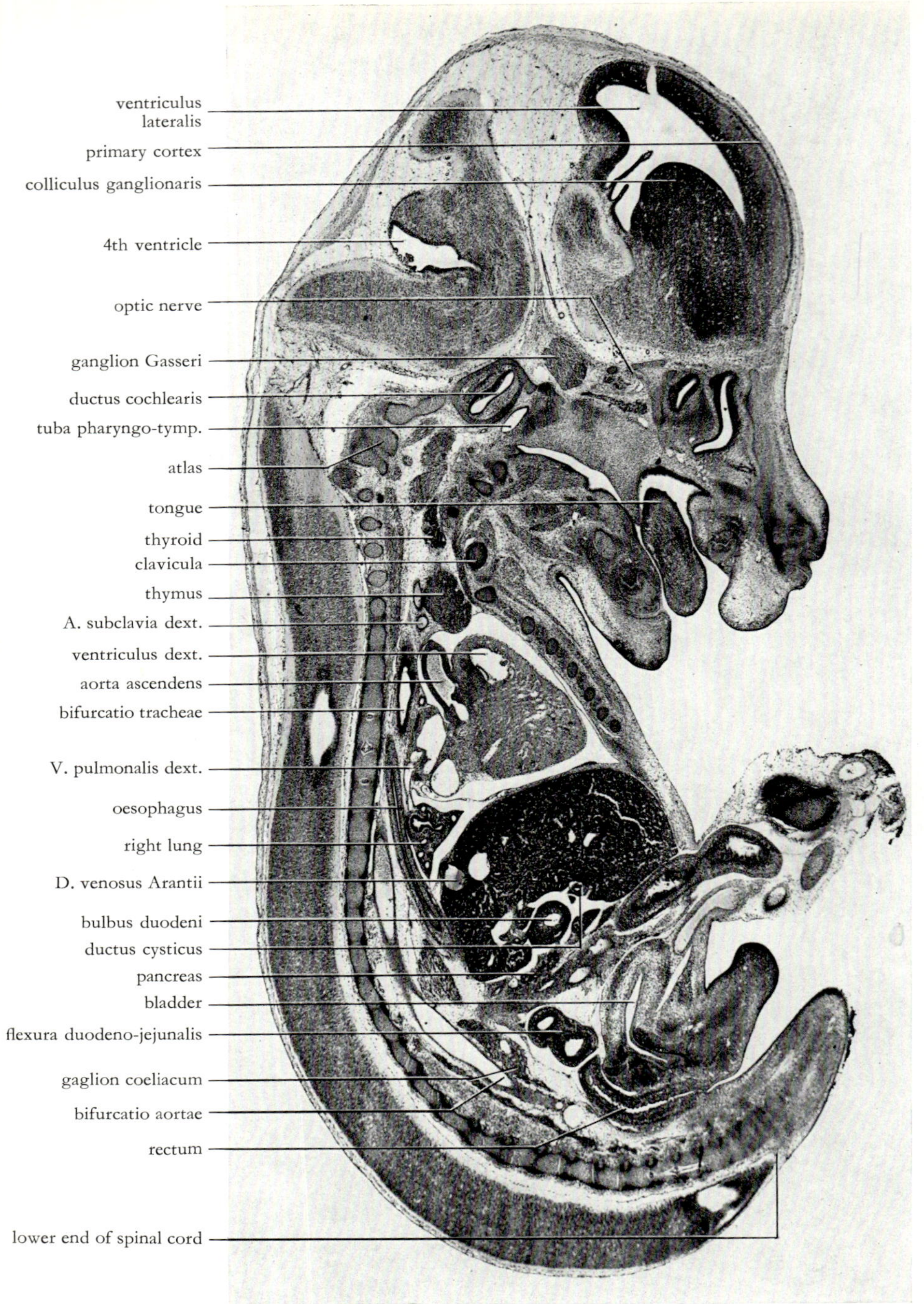

Fig. 223
Sagittal section, female.
The choroid plexus is now projecting far into the lateral ventricle. The wall of the hemisphere, which is cut here, shows peripherally a small primary *cortical layer*. The vertebral column has become chondrified. It contains relatively big notochordal segments. Some of them are visible in the thoracic region of the section (Fig. 223). At the base of the tail they are still small.
KT 643b, 14 days, 11.5 mm

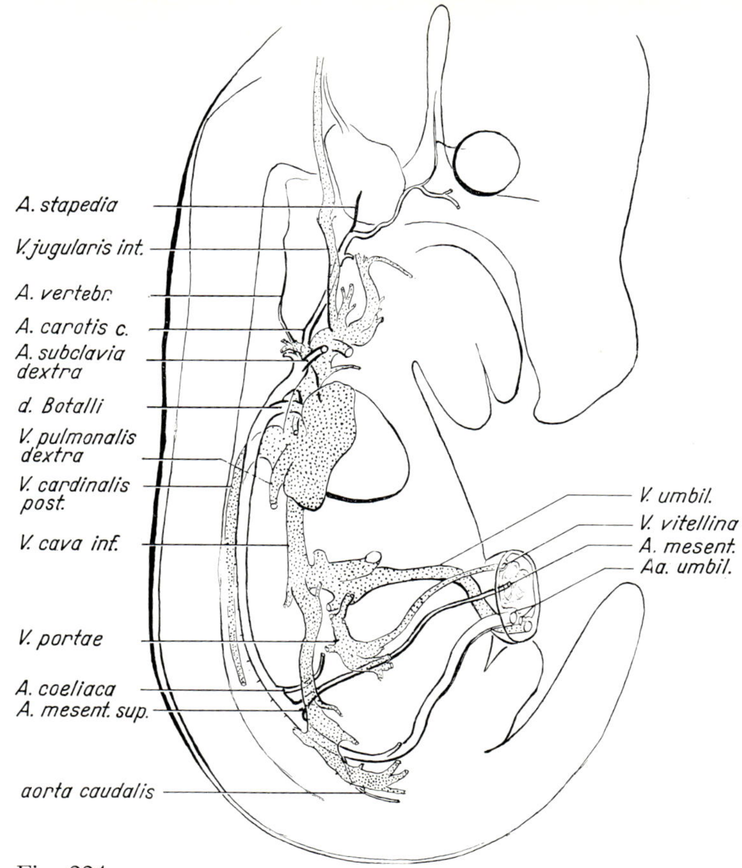

Fig. 224
Reconstruction of the vascular system, viewed from the right.
Venous system: stippled.
KT 643b, 14 days

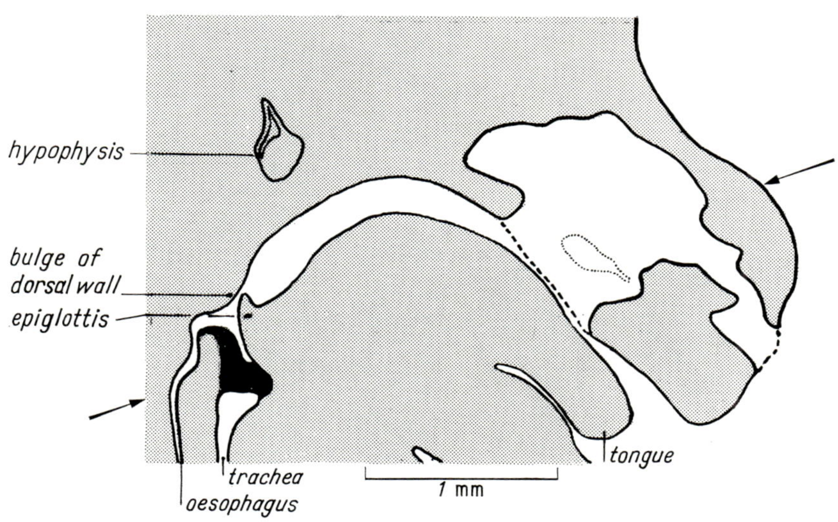

Fig. 225
Reconstruction of the nasal cavity.
Plane of section Fig. 226 indicated by *arrows*. Broken line: communication with oral cavity. Lightly stippled contour: localization of Jacobson's organ, within the nasal septum

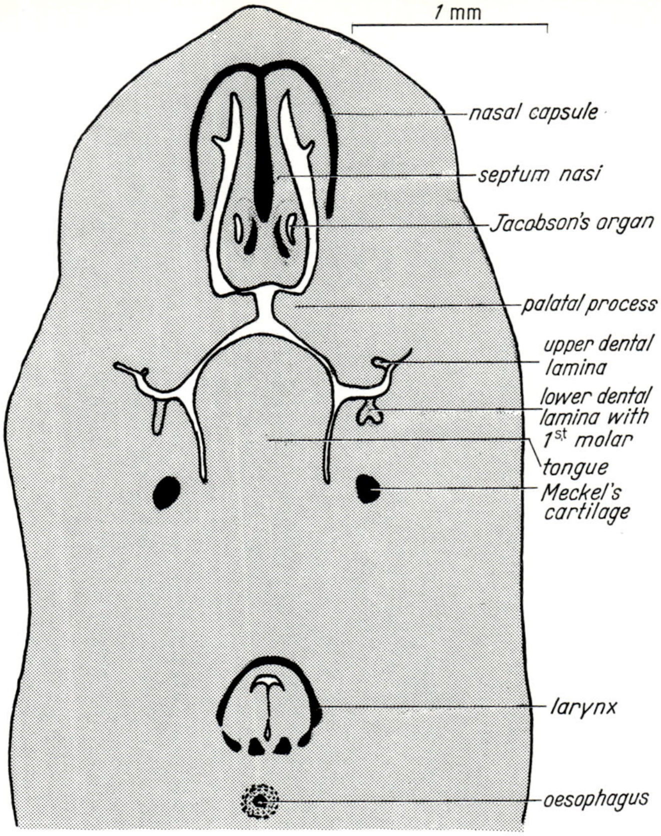

Fig. 226
Transverse section of nasal cavities (for plane of section see Fig. 225, *arrows*)

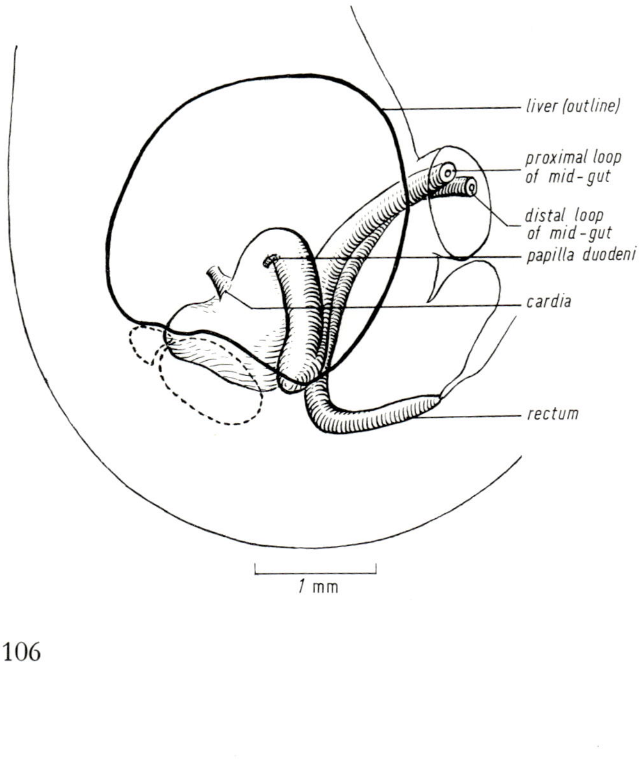

Fig. 227
Reconstruction of mid- and hindgut, viewed from the right.
Kidney and suprarenal: stippled.
KT 643b, 14 days

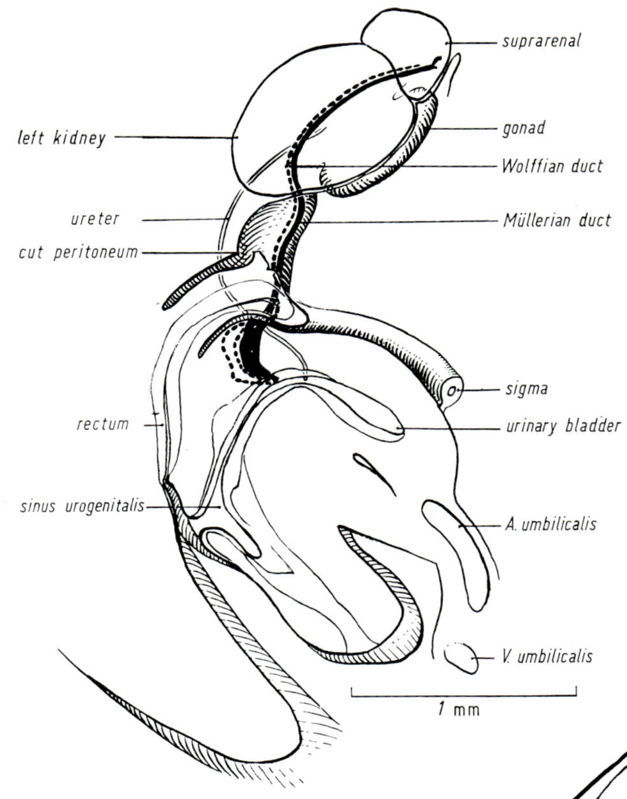

Fig. 228
Urogenital system. Reconstruction of left side.
Müllerian duct: black. Wolffian duct: broken line.
KT 643b, 14 days

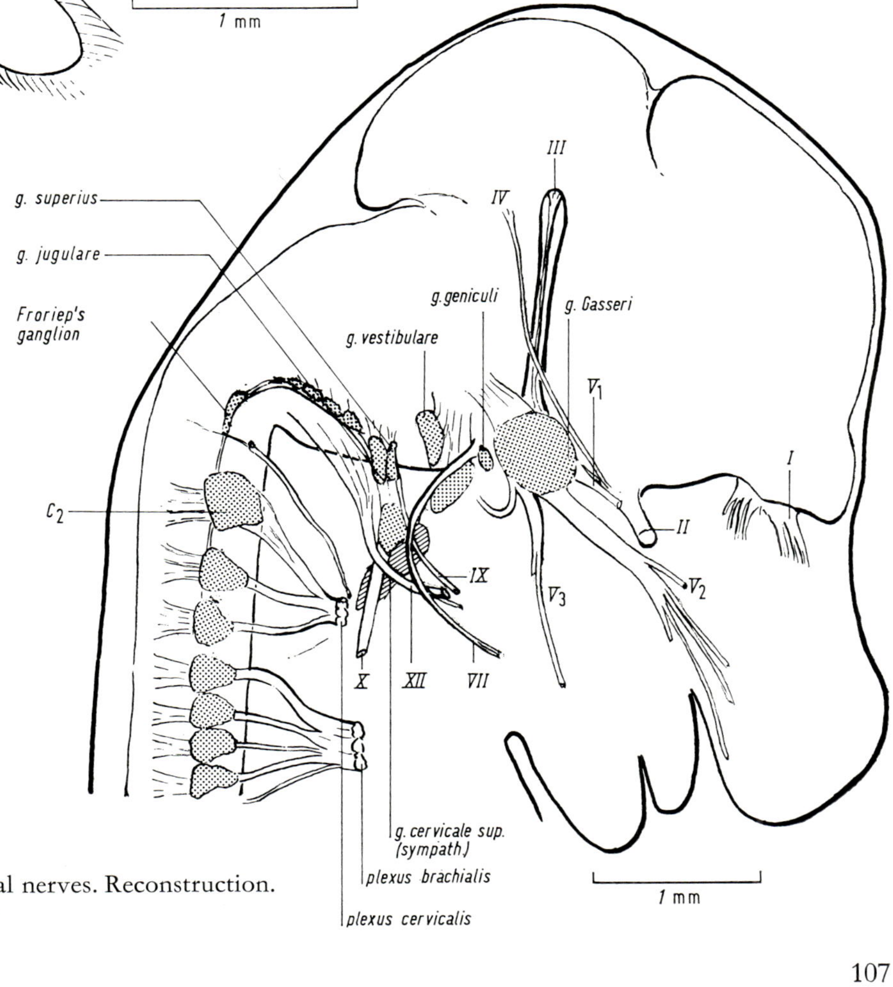

Fig. 229
Cranial and cervical nerves. Reconstruction.
KT 643b, 14 days

13 1/2 days 14 days 16 days

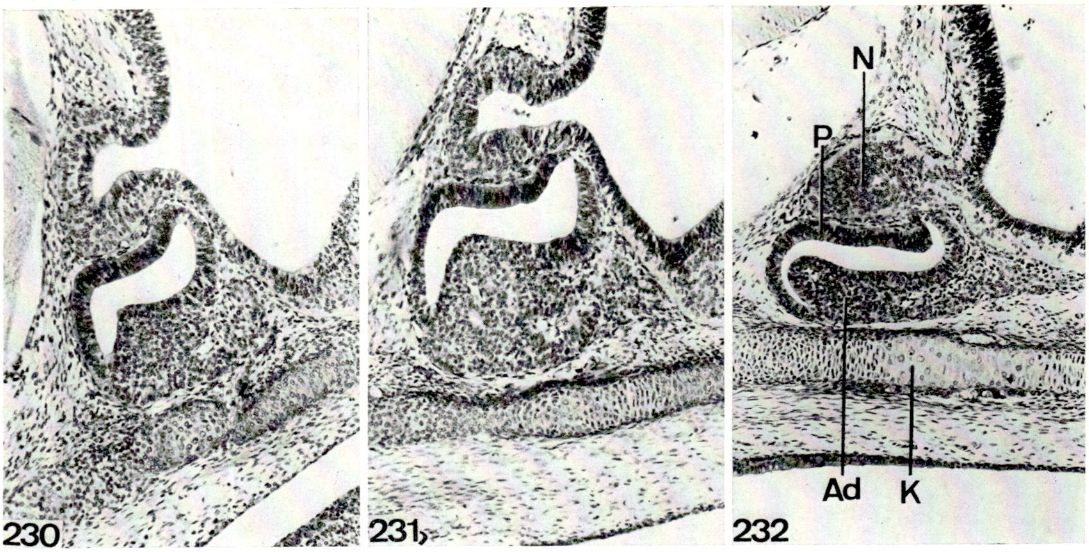

Figs. 230–232
Development of hypophysis. Sagittal sections, 13 1/2–16 days (Figs. 230 and 231: 115:1. Fig. 232: 88:1).
K Cartilaginous base of skull (sphenoid), *Ad* adenohypophysis, *N* neurohypophysis, *P* pars intermedia

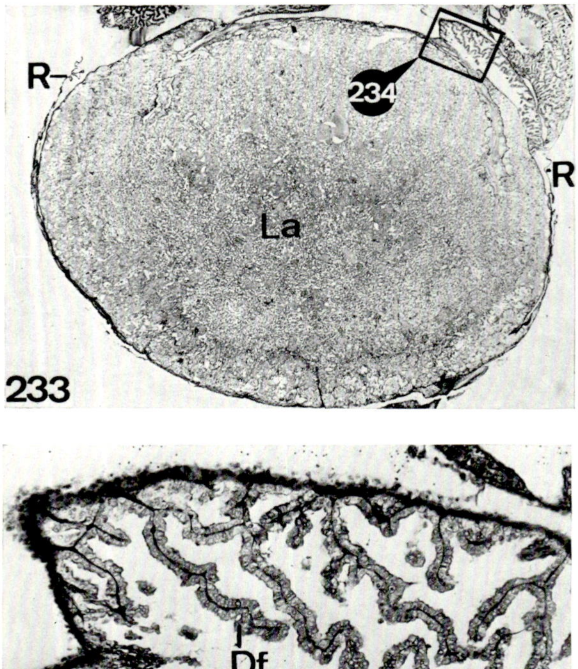

Fig. 233
Section of placenta, PAS-reaction.
La Labyrinth, *R* Reichert's membrane.
KT 1017, 14 days 2 h. 12.5:1

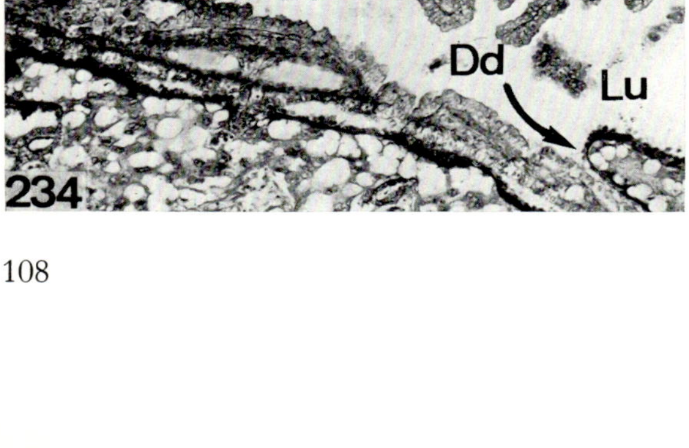

Fig. 234
Enlarged view.
Df Fold of yolk sac wall, *Lu* lumen of yolk sac, *Dd* "diverticle of yolk sac" (communication between lumen of yolk sac and intraplacental space)

Comparable to homo
2 months
(30–40 mm)

Stage 23 Toes Separate

15 Days, 12–14 mm

External Features

The most prominent feature to define this group is the separation of *toes* and "fingers" (Figs. 235, 236). They are clearly divergent and will not become parallel until much later. The *pinna* covers more than half of the external auditory meatus. In the unfixed state, the superficial veins are often distinctly visible (Fig. 235).

In fixed specimens, the discrete elevations of the hair follicles may be seen all over the body. The eylids are still open (Fig. 237). The 65 somites cannot be detected by external inspection.

Length: The length varies from 11.5 to 14 mm. In sagittal sections there are no important differences from the previous and the following stage.

Circulatory System

In the *heart*, the atrio-ventricular and semilunar valves are well developed. The stems of the coronary vessels are distinctly visible. The walls of the ventricles are smooth externally; internally, there are numerous indentations of small sinuses. The thickness of the wall is difficult to determine because the angle of section varies. The left ventricle seems to have a thicker wall than the right.

Arteries and *veins* now have the final fetal configuration, providing a larger umbilical–than yolk-sac circulation (Fig. 147). The details of the vascular system of this period will not be described.

Intestinal Tract

The oral and nasal cavities are completely separated by the palatal processes which are now fusing with the nasal septum.

The *salivary glands* are distinct glandular trees. Most branches are solid epithelial cords leading to slender excretory ducts. The initial budding occurred at 12 days.

The enamel-organs of the incisors (Fig. 233, not labeled) are in an advanced state of development. The newly arisen enamel organs of the first molars have a stellate reticulum (Fig. 226).

The cartilage of the *larynx* and of the upper trachea is well developed.

The *thyroid* is subdivided into numerous small buds. Between these buds, abundant blood vessels can be observed. Follicles have not yet formed.

The *parathyroid* is now embedded in the thyroid. The ultimobranchial body is enclosed by thyroid tissue, and is said to give rise to the parafollicular cells.

The *thymus* can be recognized as a definite lymphatic organ. It is divided into lobules which are not yet separated into medullary and cortical zones. In the center, there are many free lymphocytes. They are thought to be of mesenchymal rather than epithelial origin.

The *larynx* and *trachea* have a cartilaginous skeleton. The *lung* tissue is still rather compact, and it is more intensely vascularized than previously (compare Figs. 223 and 252).

Gut: Within the small intestine, numerous relatively thick villi have developed. In the large intestine, crypts are forming.

The *stomach* is distinctly separated into two parts. The glandular part has a tall columnar epithelium which forms tiny folds. To the left, it is joined by the nonglandular portion, which has cuboidal and stratified epithelium. The musculature is arranged in several layers.

The *spleen* contains numerous blood vessels.

The *umbilical hernia* is still present.

Urogenital Tract

The *kidneys* contain centrally placed large glomeruli, with cuboidal perivascular cells. At the periphery of the organ, there is a wide zone of metanephric blastemal tissue which borders the delicate capsule. Topographically, there is little change since 14 days (Fig. 228).

Sexual differentiation is greatly advanced.

The *ovary* has many dividing gonocytes, and they are often grouped in clusters.

In the *testis*, the solid seminiferous tubules are well differentiated. Near the surface, a condensation of cells forms the *tunica albuginea*. Diagnosis of sex is now very easy.

Central Nervous System

The *cortex* of the developing hemisphere is easily recognized at this stage. At the anterior boundary of the diencephalic roof, a small cellular area may be delimited, slightly anterior to the "Velum transversum" [158]. Some authors consider this cellular area to represent the paraphysis (Fig. 252, 16 days).

Fig. 235 Fetus of 14 days 23 h, 14 mm, life photograph. Distinct blood vessels.
KT 728. 4.7:1

Fig. 236 Different size in littermates of 14 days 20 h. Bouin fixation, 13.1 and 14.2 mm.
Nh Umbilical hernia.
KT 607 and 608. 4.5:1

Fig. 237 Eye, sagittal section, 14 days 23 h.
Lid Lower eyelid, *P* pigmented layer, *C* cornea, *G* ganglion layer, *K* nuclear layers.
KT 727. 105:1

Fig. 238 Hindfoot plate, 15 days 2 h.
Ti Tibia, *Ta* talus, *Ca* calcaneus, *Cu* cuboid, *V.* metatarsale V.
KT 1030. 40:1

Fig. 239 Forefoot plate, 15 days 2 h.
Ra Radius, *Tr* triquetrum, *Sc* scaphoid-lunatum, *II.* metacarpale II.
KT 1029. 40:1

Fig. 240 Radius, 15 days 2 h.
Ar Articulatio humero-radialis, *Ra* radius.
KT 1029. 105:1

Fig. 241 Radius of Fig. 240 under high power.
K Periostal bone. 270:1

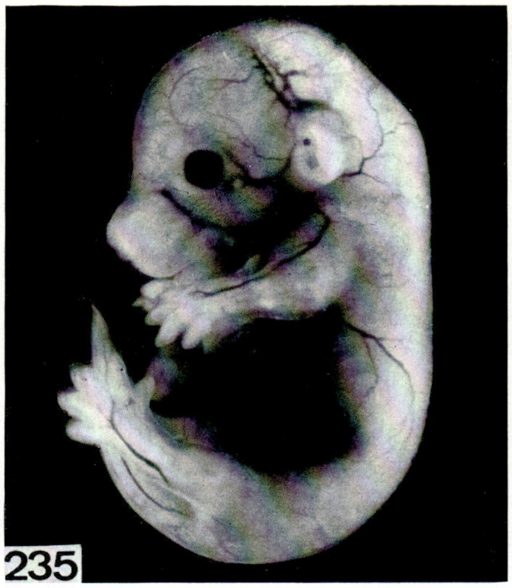

235

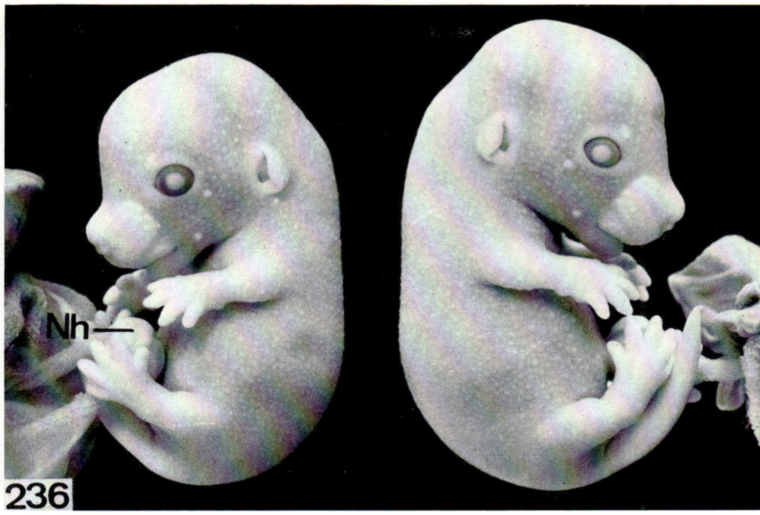

236

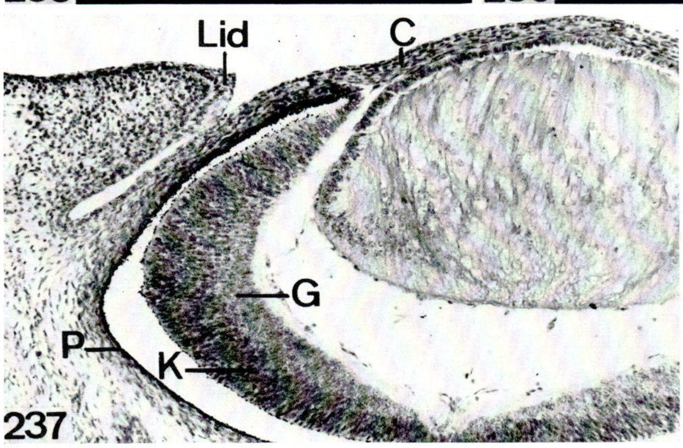

237

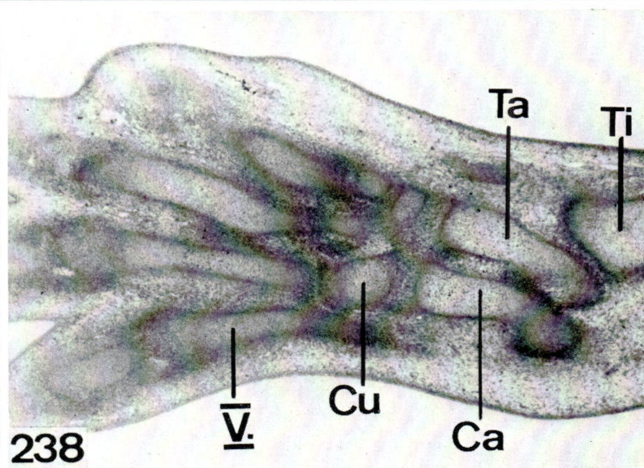

238

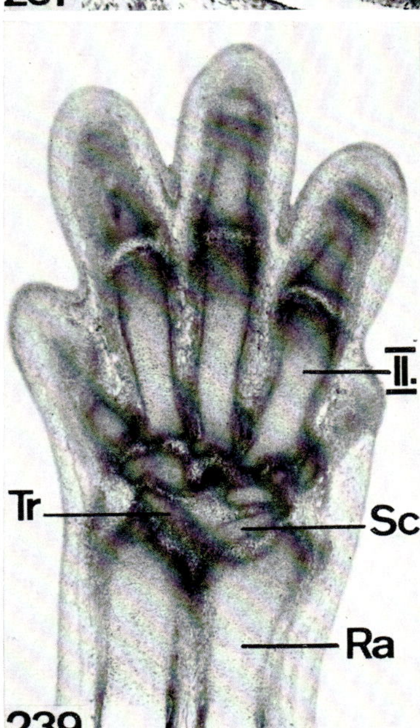

239

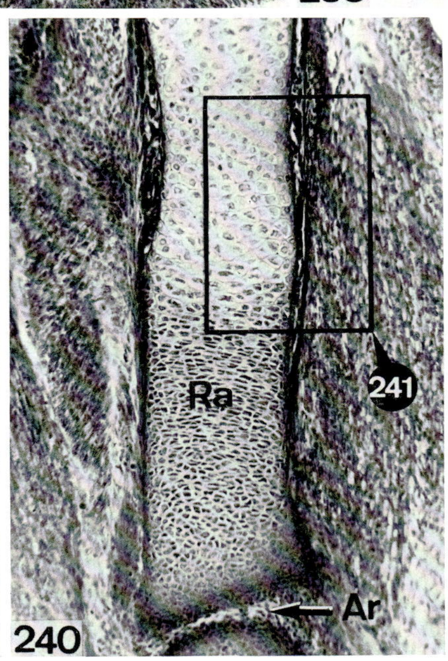

240

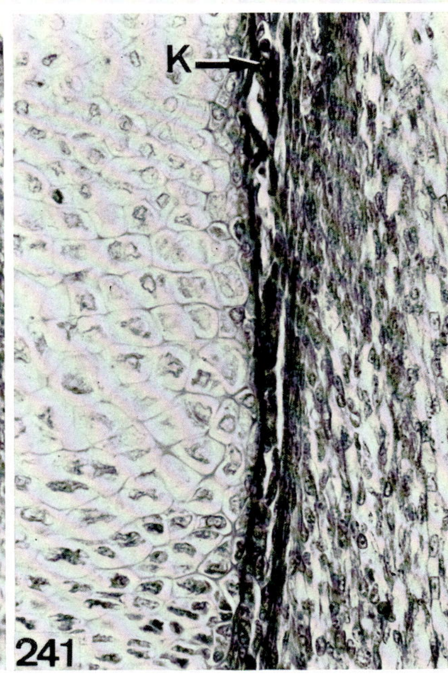

241

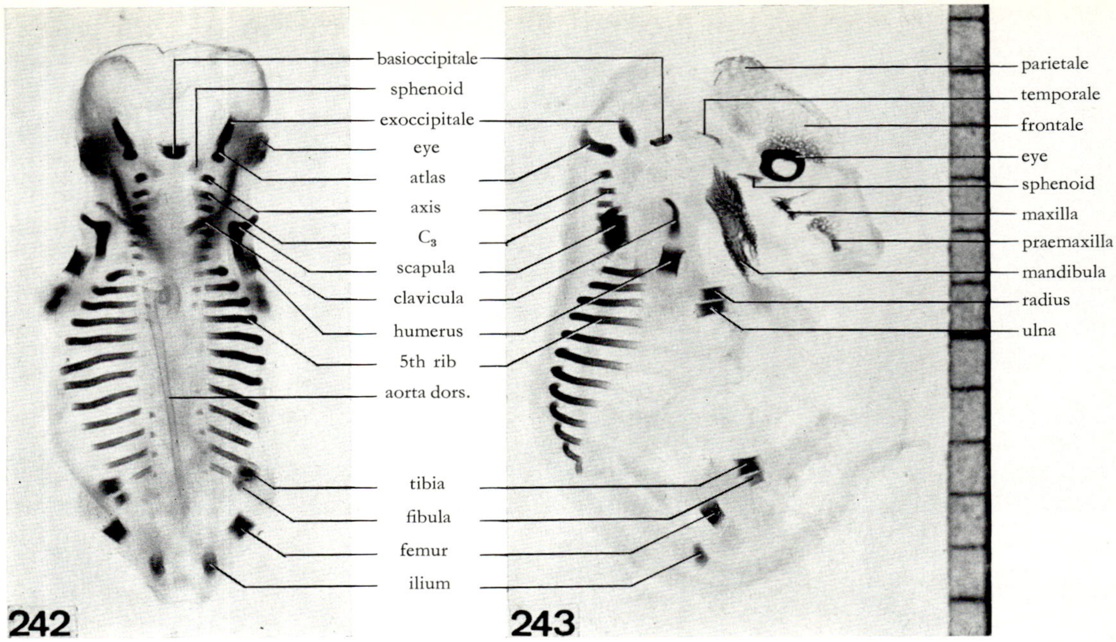

Fig. 242
Skeleton, dorsal view, 15 days 2 h.
Alizarin-cleared preparation.
KT 1032

Fig. 243
Right half of skeleton, lateral view, 15 days 2 h.
Alizarin-cleared preparation.
KT 1032

The *hypophysis* is no longer connected to the pharyngeal roof (Figs. 231–232).
The *epiphysis* (Fig. 252, 16 days) acquires a lobular shape, but still has a central lumen.
The *eye* lids are more prominent (Fig. 237). The retina near the pupillary margin still has only one layer of cells. Located adjacent to the optic nerve, the ganglion cell layer may be distinguished from the undifferentiated nuclear layers.

Skeletal System

Alizarine-stained cleared preparations show many ossification centers; some of them having arisen during the preceding stage.
In the *skull*, ossification of the *os temporale* has just started, and it is easily identified in lateral view (Fig. 243).
Ossification centers of some vertebral arches are visible, but they have not yet appeared within the vertebral bodies (Fig. 242).
In the pelvic girdle, only the ossa ilii are stainable.
During this period, the long bones of the extremities show only periostal ossification in microscopic sections (Figs. 239–241).

Material	Age	
KT 727–28	14 days 23 h	6 embryos, 13.8–14 mm + 1 resorption
KT 1029–32	14 days 2 h	Several embryos from 11.5–12.5 mm

Stage 24 Reposition of Umbilical Hernia

16 Days, 14–17 mm

External Features

In the forefoot, *fingers* 2–5 are nearly parallel to each other, while the toes of the hindfoot still diverge. The *eyelids* have fused in most specimens. An epithelial mantle, a thin transparent membrane covering the cornea, is difficult to see by external inspection (Fig. 245). The external auditory meatus is almost completely covered by the *pinna*.
The *umbilical hernia* is disappearing, and the skin is becoming wrinkled.
The anterior part of the back is completely straight; this is more apparent than in 13–15 days embryos because of the increased length. In microscopic sections, however, the anterior vertebral column has a distinct lordotic curvature (Figs. 223, 252).
Length: The length ranges, in extreme cases, between 14 and 18 mm. *Sagittal section* (Fig. 252): The primary cortex of the *brain* is thickened and the choroid plexus is larger and divided into folds and villi.
The *abdominal cavity* has enlarged, so that the intestinal loops can be repositioned as the umbilical hernia is reduced.

Circulatory System

The heart and great vessels have the final prenatal configuration (Fig. 252).
The superficial veins are easily recognizable through the thin transparent skin (Figs. 245, 246).

Intestinal Tract

A long slender *ductus nasopharyngeus* develops by further downward growth of the fused palatal processes. In this way, the epipharynx extends as far back as the prominent *epiglottis* (Fig. 252). The same sagittal section shows the anlage of the *incisors*.
The lip furrow (vestibulum oris) is still a solid plug of epithelial cells which will later split.
The *thymus* increases strikingly in volume, surpassing considerably the neighboring thyroid (Fig. 252). The *thyroid* does not yet have follicles. Nevertheless histologically it can be clearly distinguished from the parathyroids (Fig. 249).
The structure of the *lung* has not changed much since the previous stage.
The *stomach* is rapidly enlarging. The nonglandular portion, illustrated in Fig. 247, is lined by a multilayered cuboidal, sometimes flattened (squamous) epithelium.
The *spleen* (Fig. 247) contains distinct arteries and veins, separated by a rather compact tissue composed of various types of cells. Sometimes phagocytes can be recognized (Fig. 248) which seem to take up nuclei of red blood cells.
The *small intestine* now has longer, but still rather thick villi covered by columnar epithelium.
The large intestine is forming crypts.
In the *liver*, blood cell production is increasing. Externally, the final lobation is apparent [81].
The *pancreas* has finely branched, glandular trees, with distinct lumina, and pancreatic islets

are budding. The typical β-cells cannot be seen with the light microscope until the 17th or 18th day [139].
The *suprarenals* appear as a regular network of strands composed of eosinophilic cells separated by wide capillaries. Scattered clusters of small medullary cells may also be seen.

Urogenital Tract

The *kidneys* still have a large peripheral metanephric blastema (Fig. 247). Near the center, many glomeruli are well developed.
In the *testes* distinct interstitial cells can now be recognized.

Central Nervous System

The primary cerebral cortex has enlarged considerably. In the diencephalon, the differentiation of the hypophysis and of the pineal gland is progressing. The epiphysis still has a central lumen, but the wall is markedly thickened (in Fig. 252 the wall is sectioned tangentially). Anterior to the velum transversum, a small area represents the paraphysis (GRAUMANN [158]).

Fig. 244 Eye, horizontal section, 16 days, 15 mm.
Lid Lower eyelid, *C* cornea, *L* lens, *G* ganglion layer of retina.
KT 646. 105:1

Fig. 245 Fetus of 16 days, life photograph, 15 mm.
KT 646

Fig. 246 Diagram of Fig. 245.
V.u. Vena umbilicalis, *V.v.* vena vitellina, *Nh* umbilical hernia (partially repositioned), *P* placenta, *O* pinna, *V.t.* vena temporalis superficialis, *V.c.* vena cephalica, *V.th.* vena thoracica lateralis, *V.m.* venae metatarseae dorsales

Fig. 247 Cross section, at the level of the spleen, 16 days.
Ma Stomach, *Md* mesogastrium dorsale, *Ni* left kidney, *RM* spinal cord, *Ar* rudiment of vertebral articulation.
KT 646. 40:1

Fig. 248 Enlarged view of spleen in Fig. 247, with phagocyte (*arrow*) containing presumably nuclei of erythroblasts.
A Artery. 550:1

Fig. 249 Cross section, level of larynx, 16 days.
Ly Lymph vessel, *X.* nervus vagus, *Vj* vena jugularis interna, *Ac* arteria carotis communis, *Th* thyreoidea (thyroid gland), *Pt* parathyroid, *S* cartilago thyreoidea, *La* cavum laryngis.
KT 646. 105:1

Fig. 250 Sagittal section through 5th thoracic vertebra.
RM Spinal cord, *Ch* notochordal segment, *Cs* notochordal sheath, situated dorsally in calcified center of cartilage.
KT 646, 15 days. 105:1

Fig. 251 Sagittal section through axis, 16 days.
Cs Notochordal sheath, on clivus; *A* anterior arch of atlas; *Af* anulus fibrosus; *C3* body of 3rd cervical vertebra.
KT 646. 105:1

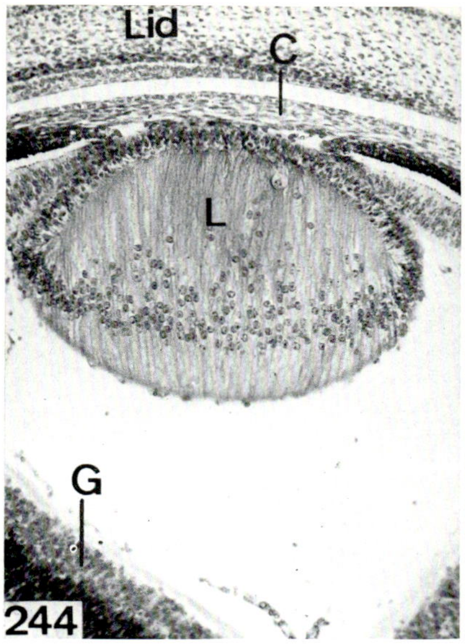

244

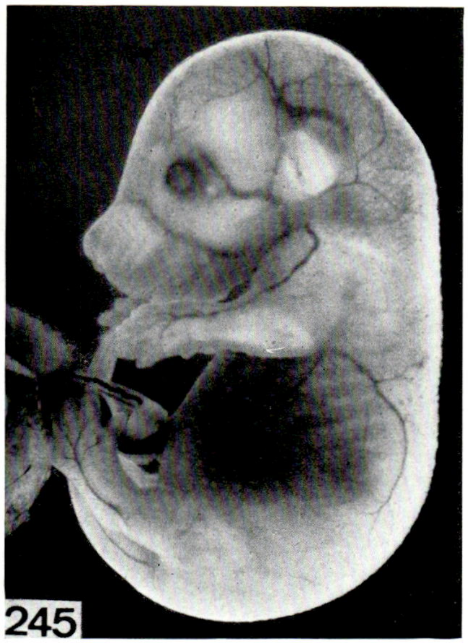
245

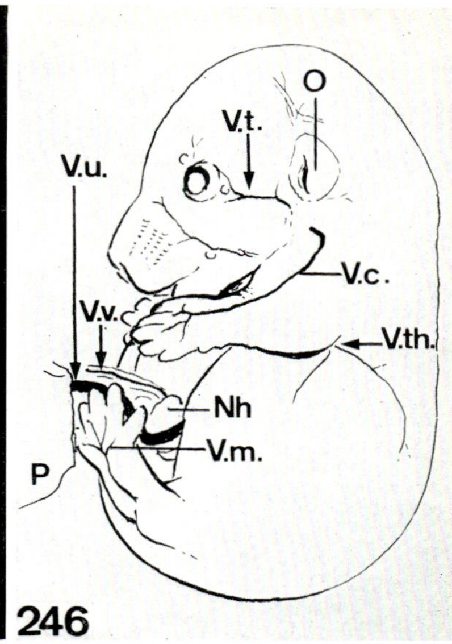

246

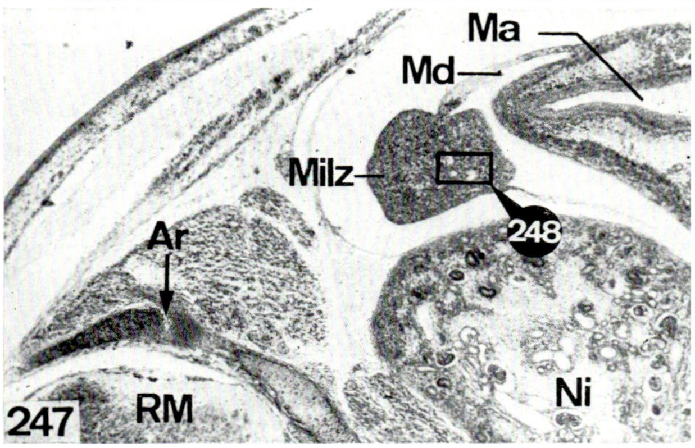

247

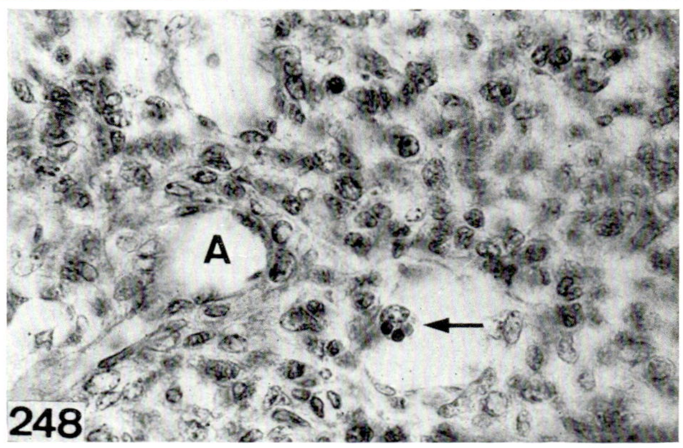

248

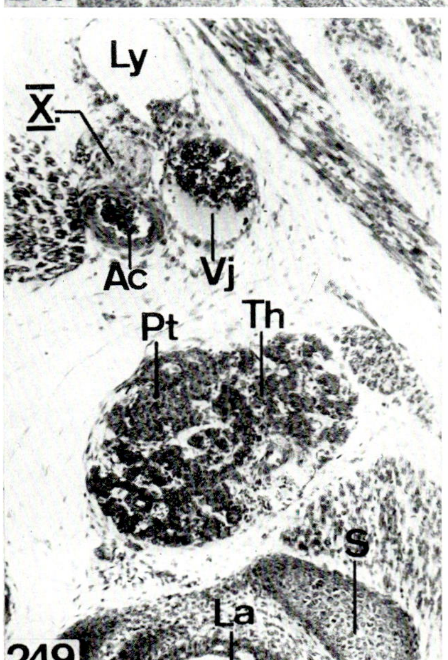

249

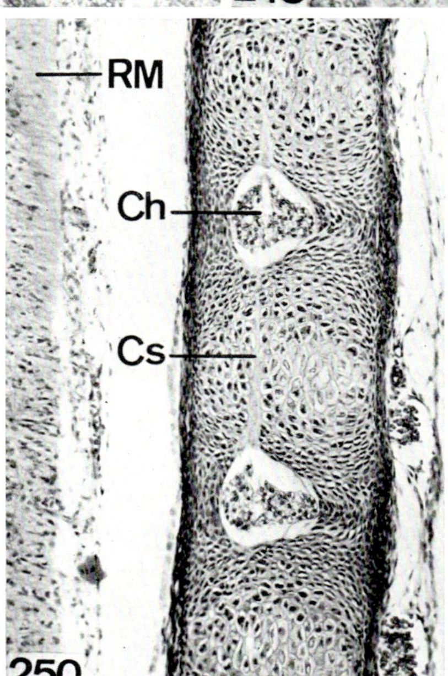

250

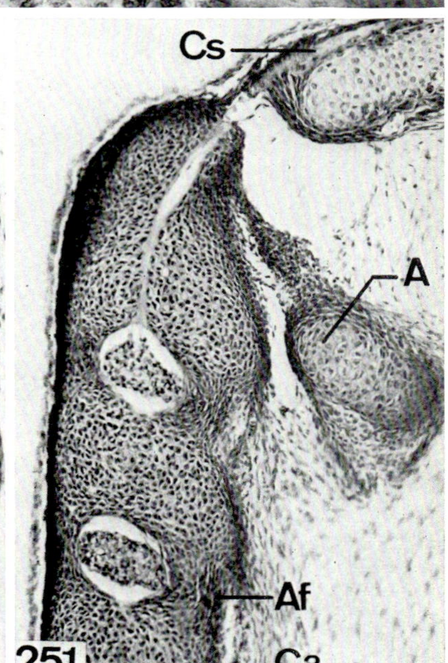

251

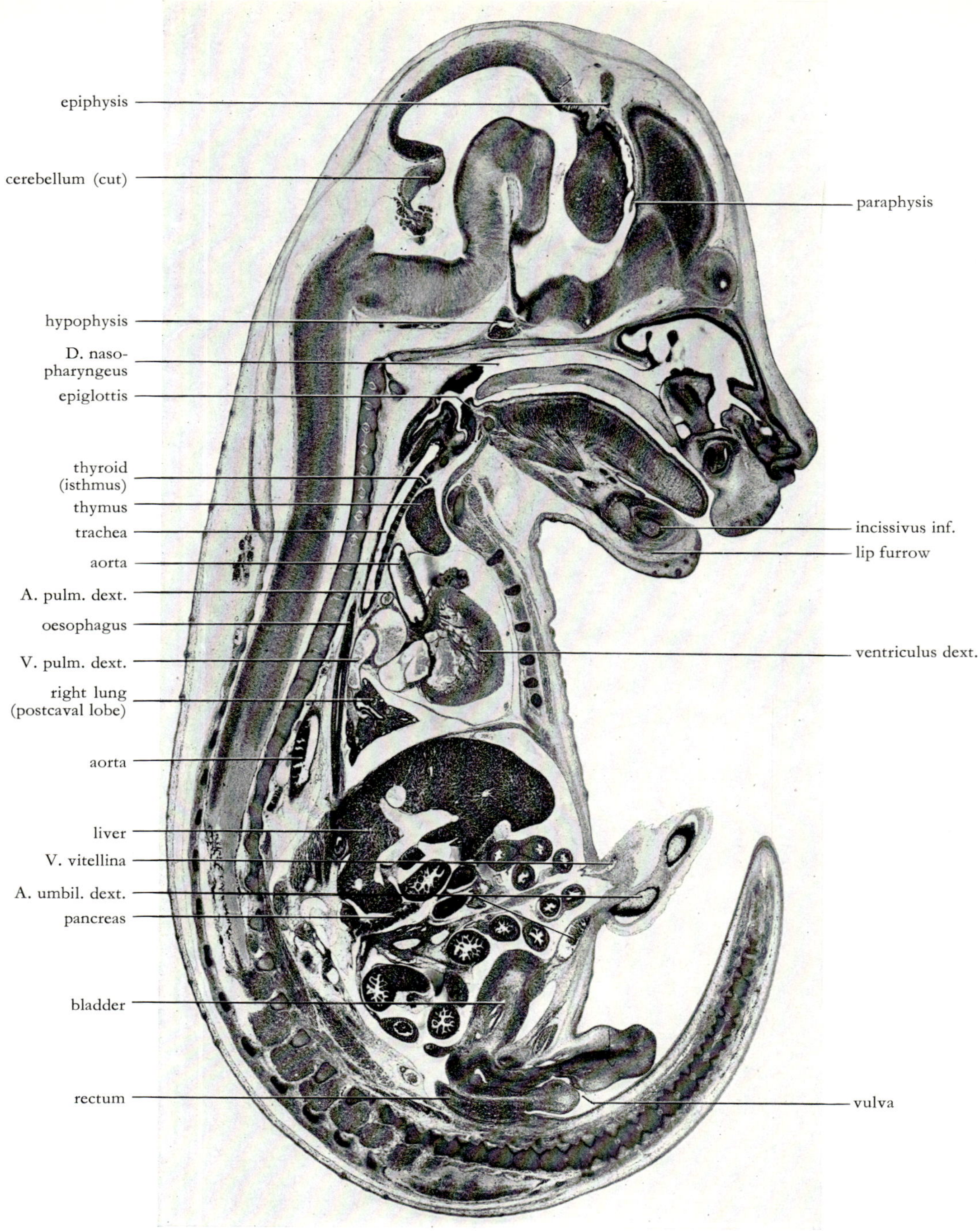

Fig. 252
Sagittal section. Female fetus, 16 days, 14.6 mm.
KT 646

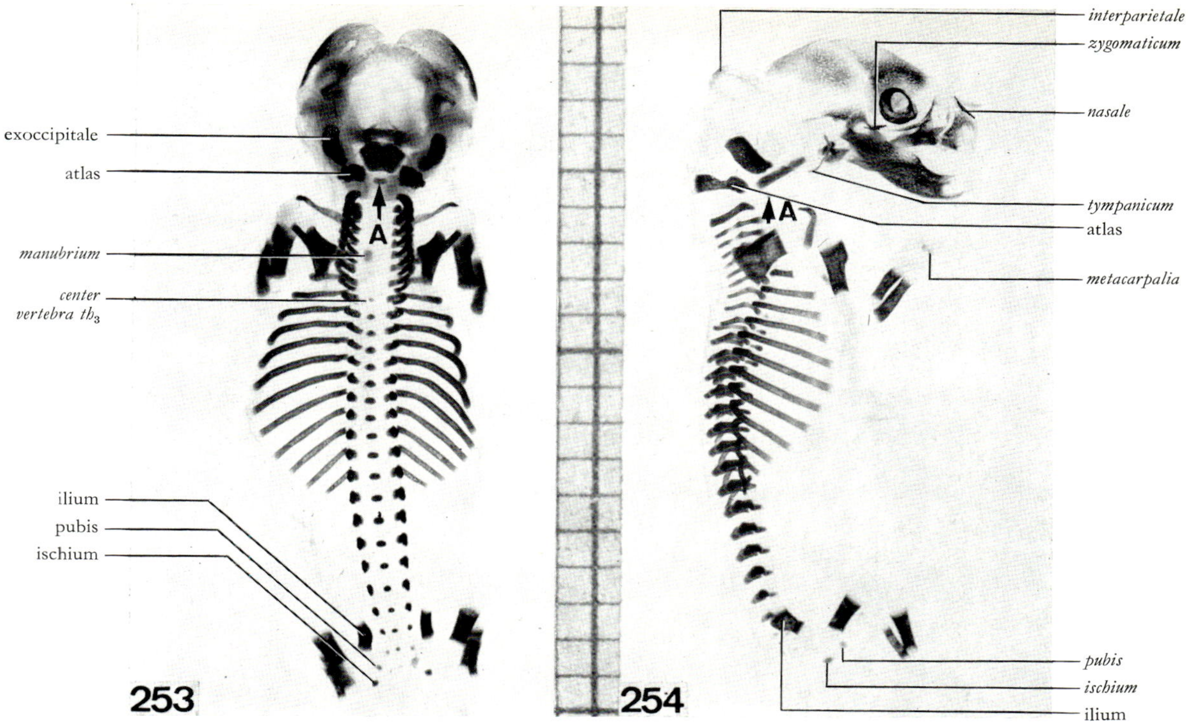

Figs. 253–254
Alizarin-cleared preparations. Fetuses of 16 days. *New skeletal elements in italics* (in the previous stage, Figs. 242–243, not yet visible).
Arrows : ossification center of anterior arch of atlas.

A thin transparent epithelial membrane is growing over the cornea of the *eye*. The iris and the corpus ciliare cannot yet be distinguished. The anterior chamber of the eye extends beyond the pupillary margin (Fig. 244).
The *retina* is not much more differentiated than in the 15-day stage.
The labyrinth has a well developed cartilaginous capsule which encloses the semicircular ducts and the ductus cochlearis. The sensory epithelia are thickened, but not well differentiated.

Skeletal System

Most of the vertebral bodies contain initial ossification centers, i.e. calcium deposits within the cartilage, as shown in Fig. 250. Photographs of sagittal sections through the 5th thoracic vertebra will illustrate, in later stages, progressing ossification.
A salient feature of this developmental stage is the appearance of ossification centers in the anterior arch of the atlas and in other skeletal elements whose labels are underlined in Figs. 253–254.

Material	Age	
KT 646–47	16 days 2 h	7 fetuses, 14.0–15.2 mm
KT 1046–48	16 days 4 h	5 fetuses, 16–18 mm + 1 resorption

Stage 25 Fingers and Toes Joined Together

17 Days, 17–20 mm

External Features

The skin is wrinkled and thickened, and the subcutaneous veins are no longer distinctly visible (compare Figs. 256 and 245). The eyelids are fused and thickened. The fingers and toes are parallel. The umbilical hernia has disappeared in all cases examined.
Length: There is considerable variation in length due to different degrees of curvature of the fetuses: 16.5–20 mm in the unfixed state.

Circulatory System

The pattern of the superficial veins has not changed since the previous stage, but is more difficult to recognize.

Intestinal Tract

The shape of the oral and nasal cavities has not changed much.
Endocrine derivatives: The *thyroid* is a bi-lobed organ, with a narrow isthmus. The solid, branched epithelial cords are beginning to form small well vascularized follicles.
The *parathyroids* are forming compact clusters of cells on each side, which join the posterior margins of the thyroid lobes.

Fig. 255 Living fetus, frontal view, 17 days, 20 mm.
KT 908

Fig. 256 Living fetus, viewed from the left, 17 days, 20 mm.
KT 908

Fig. 257 Living fetus, viewed from the right. Same developmental stage, but one day older, 18 days, 20 mm.
KT 657. 4:1

Fig. 258 Eye, horizontal section, enlarged view of Fig. 260.
C Cornea, *I* iris rudiment, *Cc* corpus ciliare, *G* ganglion layer of retina, *Pl* punctum lacrimale (on inner surface of eye lid).
KT 724, 17 days. 105:1

Fig. 259 Pineal gland, sagittal section, 17 days.
V Vena cerebri interna.
KT 724. 105:1

Fig. 260 Low power view of eye, with nervus opticus and vicinity.
K Caput mandibulae, *gG* ganglion Gasseri, *N* nasal cavity.
KT 724, 17 days. 27:1

Fig. 261 Sagittal section through thoracic vertebrae 4 and 5 (Th_4 and Th_5) +/+ fetus, 17 days.
Cs Notochord sheath, *D* dura mater spinalis. 105:1

Fig. 262 Sagittal section through axis, 17 days.
A Anterior arch of atlas, *Af* anulus fibrosus C_2/C_3, *Lt* ligamentum transversum atlantis, *Cs* notochord sheath.
KT 908. 100:1

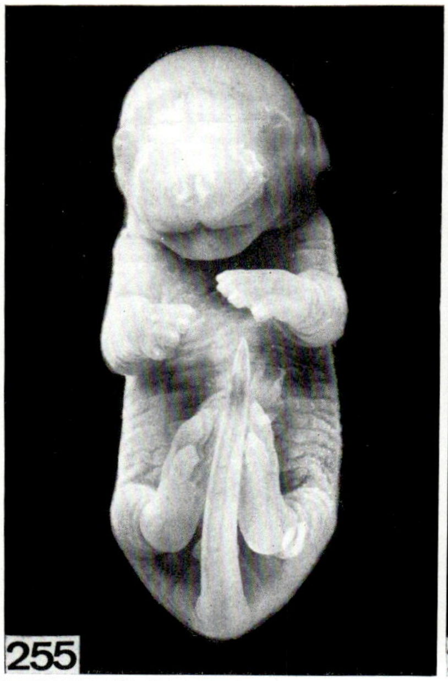
255

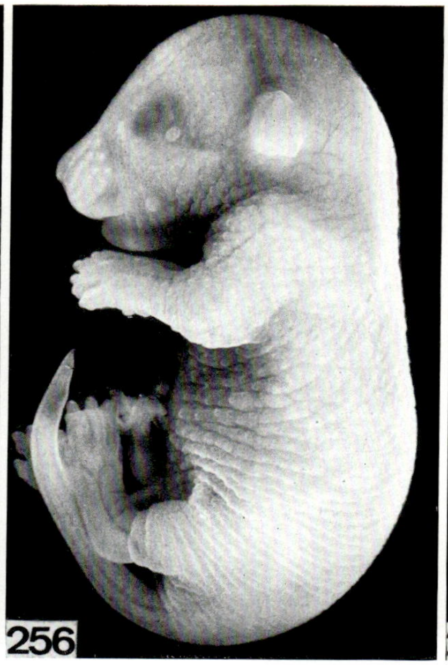
256

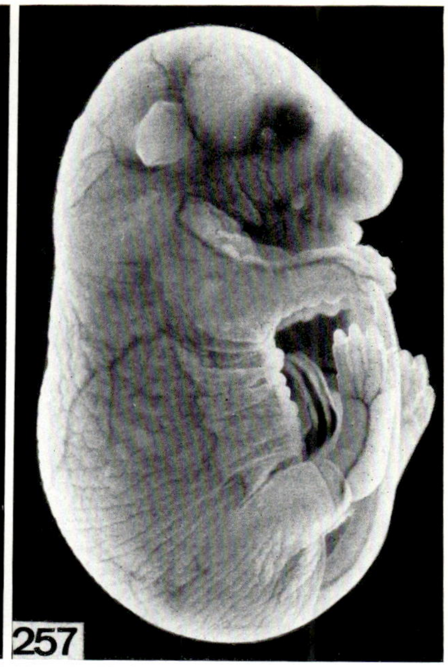
257

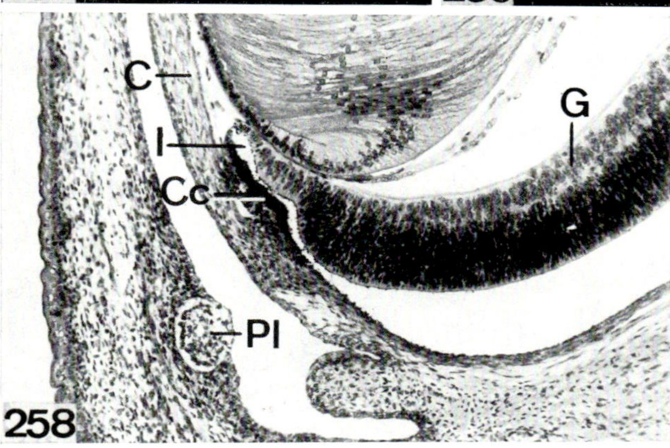
C
I
Cc
G
Pl
258

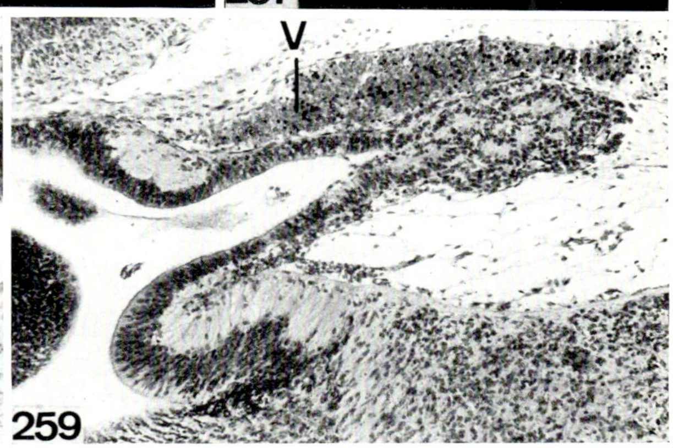
V
259

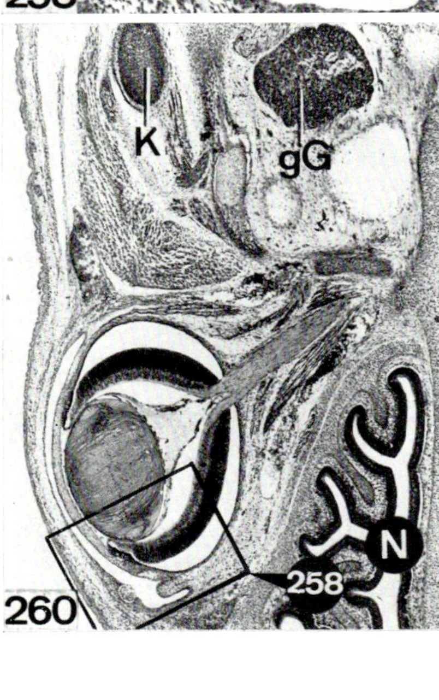
K
gG
N
258
260

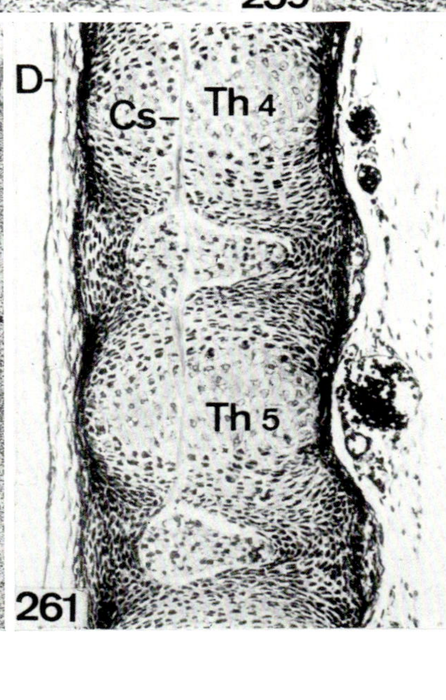
D
Cs
Th 4
Th 5
261

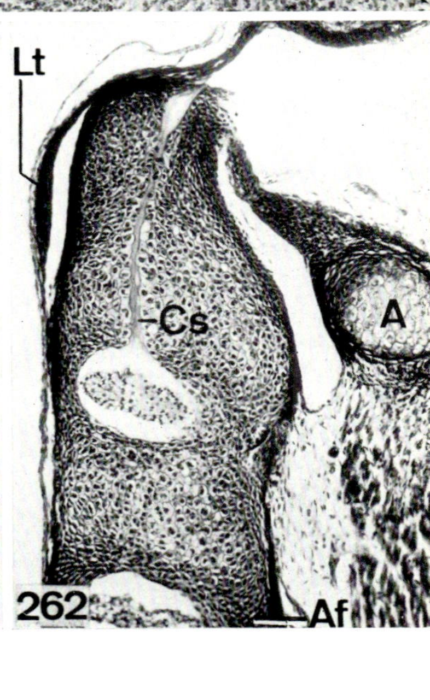
Lt
Cs
A
Af
262

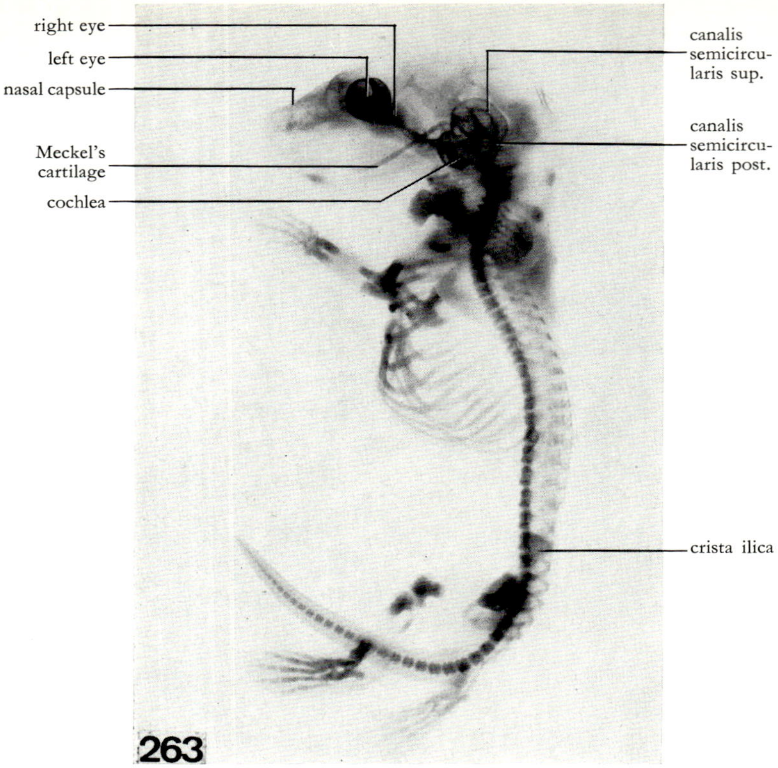

Fig. 263
Cartilaginous skeleton with 17 days. Cleared preparation, Methylgreen. KT 776

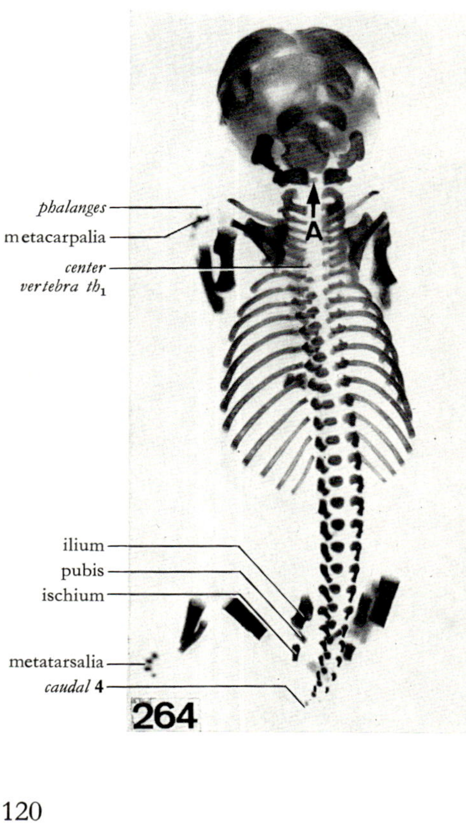

Fig. 264
Alizarin cleared preparation, 17 days 23 h. Dorsal view. *New bones in italics. A (arrow)*: ossification center in anterior arch of atlas.
KT 1028

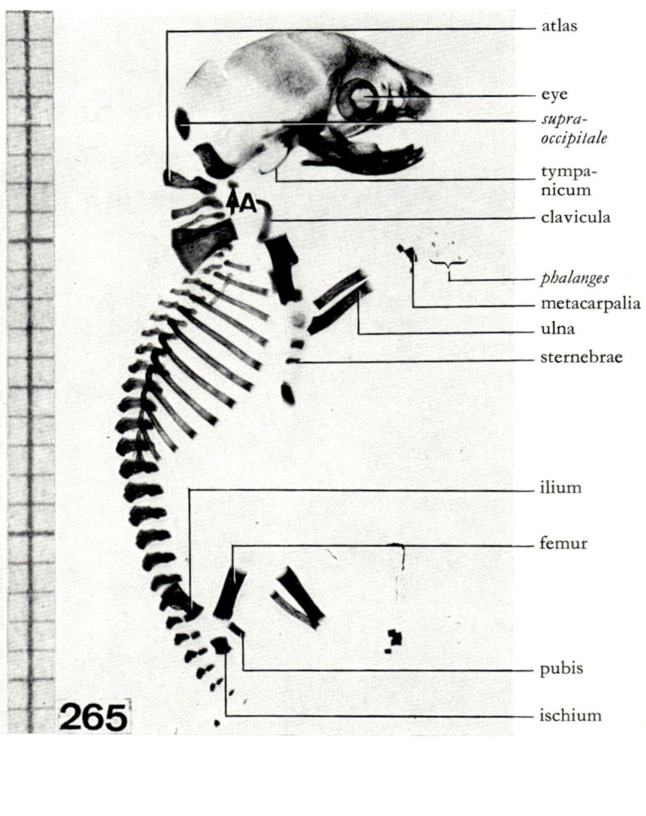

Fig. 265
Right half of skeleton, same specimen as shown in Fig. 264, with millimeter scale.

The *thymus* is now a large organ, separated into two bilaterally situated main lobes containing numerous small blood vessels. Sometimes aggregations of reticulum may be seen, but there are no Hassal's corpuscules.

The structure of the *lungs* appears less compact now. The respiratory pathways are extending peripherally. The developing alveolar ducts are lined by cuboidal epithelium.

The *intestines* are quite similar to the previous stage.

The *spleen* is elongating rapidly. It contains, for the first time, lymphocytes.

The *adrenal glands* have a more continuous medullary region consisting of small pale cells, which differ sharply from the large, intensively stained cells of the cortex.

The *testes* have well formed tunicae albugineae. Numerous scattered interstitial cells are recognizable singly or in groups.

The *ovaries* contain many oocytes, often grouped together. They have entered meiotic prophase.

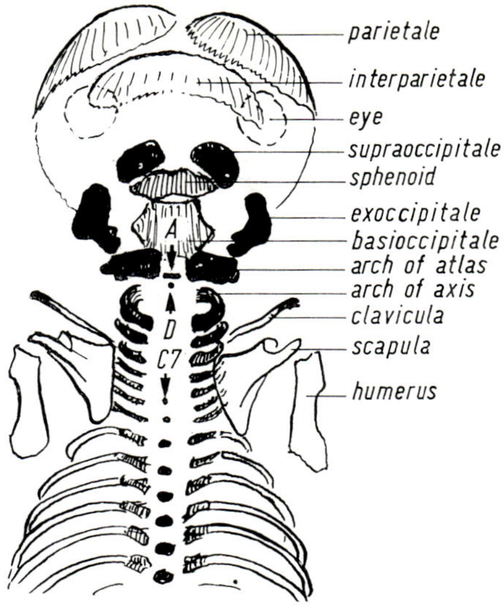

Fig. 266
Dorsal view of alizarin cleared preparation, cranial part, 17 days.
D Ossification center in basis of dens, newly arisen; *A* anterior arch of atlas; *C7* ossifacion center in body of 7th cervical vertebra.
KT 1028

Central Nervous System

Because of the gland-like budding of its epithelium (Fig. 259), the distal part of the lumen of the epiphysis is disappearing.

The eyelids are closed. The ciliary body is delineated (Fig. 258).

The capsule of the *labyrinth* has assumed its definite shape (Fig. 263).

Skeletal System

The skeletal system is best studied in cleared specimens (Figs. 264, 265). Labels of new bones, which now stain with alizarin red S, are underlined in the legend, i.e. phalanges etc. It should be pointed out that calcified cartilage may also be stained and microscopically it is different from bone. Histologic sections of some vertebrae are illustrated in Figs. 261 and 262.

The ossification of the vertebral bodies begins in the lower thoracic region and has now proceeded forward to the first thoracic vertebra. The anterior arch of the atlas has already begun to ossify, and the basal ossification center of the dens is also developing at this early stage. The centrum of the dens axis is just appearing (Fig. 266). The centrum of the axis body will appear soon.

Material	Age	
KT 724–25	17 days	8 fetuses, 17–17.5 mm
KT 908	17 days	3 fetuses, 18–20 mm
KT 1052	17 days 2 h	8 fetuses, 18–20 mm
KT 1049–50	17 days 3 h	7 fetuses, 16.5–19.5 mm

Stage 26 Long Whiskers

18 Days, 19.5–22.5 mm

External Features

The whiskers, visible at 17 days as short filaments (Fig. 255), are now definitely longer. The skin is thickened and markedly wrinkled. The eyes are barely visible through the closed eyelids.

Length: Because of different degrees of curvature, the length varies from 18 to 23 mm.

Circulatory System

The final prenatal configuration of the circulation system was established at 16–17 days.

Intestinal Tract

In the *oral cavity*, the rate of development of the first molars is remarkable. The stellate reticulum is enlarged and the outer enamel organ epithelium is in close contact with numerous capillaries. The ameloblastic layer, composed of tall cylindrical cells, has the form of the future crown. In Fig. 270, the enamel organ is separated from the surrounding loose connective tissue by artificial shrinkage. The second molar is about 2 days later in development than the first.

Derivatives of the Intestinal Tract

In the *thyroid* the number of colloid-filled follicles has increased considerably. Between the numerous lobuli, there is loose connective tissue containing many blood-vessels. The secretory activity seems to be intensive.

Both *parathyroids* may be recognized as rather compact, spherical cell clusters, closely adhered to the thyroid. They are situated on each side somewhat below the middle of the lobes. Many blood-vessels have grown between the parenchymal cords.

The *pancreatic islands* are now well differentiated.

The *thymus* has enlarged further (Fig. 275). There is no clear boundary between the medulla and cortex. In the medulla, extensive aggregations of large clear reticulum cells are visible. There are numerous blood-vessels.

Respiratory System

The structure of the *lungs* has changed completely: one could speak of an "alveolar explosion". There is a sudden development of large alveolar ducts and of sac-like primitive alveoli lined by cuboidal epithelium. The resulting loose structure of the lung tissue is visible with low magnification.

The *larynx* and *trachea* have completed their prenatal differentiation.

Abdominal Viscera

There have been no marked changes in the abdominal viscera since the preceding stage. The *spleen* has elongated, is rich in cells, and exhibits signs of intensive blood formation.
There is still active hemopoiesis in the *liver*.
In the *stomach*, the subdivision into glandular and non-glandular part is clearly visible. In the *gut*, the characteristic relief of the lining has developed.

Urogenital Tract

The *kidneys* are about the same as in the preceding age group.
The *glands* of the male genital excretory ducts are now very conspicuous, and the *seminal vesicles* (Fig. 275) contain distinct lumina. The lumen of the bulbo-urethral gland (Cowper's gland) is just forming and it opens into the urethra together with the recessus bulbosus.
The *prostate gland* is appearing as short, solid cords of epithelial cells budding from the pars prostatica urethrae.
The tunica albuginea of the *testis* has increased in thickness. The seminiferous tubules are still solid and are connected with the canalicular system of the rete testis (Fig. 276). The gonocytes have stopped dividing [118]. Between the tubuli contorti, there are numerous interstitial cells.
Ovaries (Figs. 278, 279): The ovaries are closely attached to the lower poles of the kidneys. The bursa ovarica almost completely surrounds the ovary. In the mesosalpinx, many mesonephric tubules can be recognized. Some oocytes are being encircled by flat epithelial cells which form the *primary follicles*.

Central Nervous System

The rudiment of the *pineal gland* has been transformed into a stalked, glandular organ.
Within the *hypophysis*, the original lumen of Rathke's pouch is narrowed (Fig. 271). The adenohypophysis assumes the appearance of a typical endocrine gland.
In the *brain*, the olfactory lobe is well developed.
In the *eye*, the iris and corpus ciliare are well formed (Fig. 272).

Skeletal System

Skull: The bilaterally situated ossification centers of the supraoccipitale have fused (compare Figs. 281 and 266).
Vertebral column: The vertebral bodies C_3–C_5 still lack ossification centers.
Extremities: Ossification of the extremities is proceeding rapidly (Fig. 280).
Adipose tissue is abundant in the subcutaneous tissue of the neck. It probably is a source of energy for the postnatal period, and has been called "multilocular adipose tissue" ("M.A.T.", Fig. 275).

Material	Age	
KT 656–57	17 days 21 h	5 fetuses, 18.5–20.5 mm
KT 1010–11	17 days 23 h	6 fetuses, 18.0–20.0 mm
KT 1023–28	17 days 23 h	6 fetuses, 19.0–21.0 mm
KT 1041–42	18 days 4 h	6 fetuses, 23 mm

Fig. 267 Fetus of 17 days 23 h, 22.2 mm. Formalin fixation.
KT 1027

Fig. 268 Same fetus viewed from the right. 3:1

Fig. 269 Alizarin-cleared preparation, 17 days 23 h.
Explanation see Fig. 280.
KT 1011. 3.5:1

Fig. 270 Cross section through head, level of eyes, 17 days 23 h.
CN Cavum nasi, *DN* ductus naso-pharyngeus (compare Figs. 252 and 225), *Olf* lobus olfactorius, *Sp* stellate reticulum (molar). 23:1

Fig. 271 Hypophysis, sagittal section, 18 days 4 h.
N Neurohypophysis (posterior lobe), *P* pars intermedia, *Ad* adenohypophysis (anterior lobe), *T* pars tuberalis.
KT 1041/3. 130:1

Fig. 272 Enlarged view of eye in Fig. 270.
Cc Corpus ciliare, *I* iris, *Ln* suture of eyelids. 60:1

Fig. 273 Sagittal section through 5th thoracic vertebra, 18 days 4 h.
Cs Notochordal sheath, *P* periostal bone, ventral part.
KT 1041/3. 105:1

Fig. 274 Sagittal section through axis, 17 days 23 h.
A Arcus anterior atlantis, *D* dura mater cerebri (above clivus), *De* ossification center in basis of dens, C_3 body of 3rd cervical vertebra.
KT 1010. 55:1

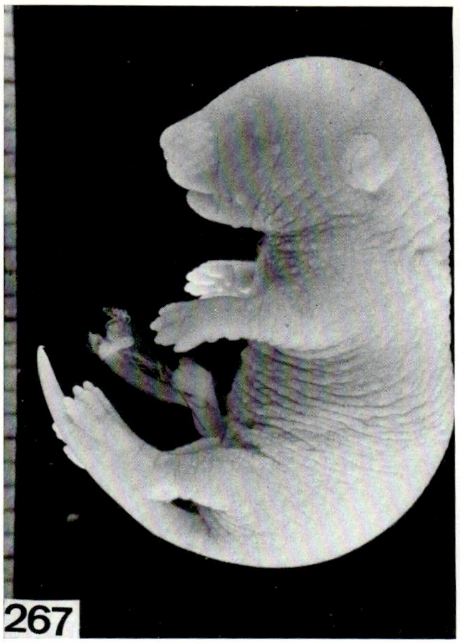

267

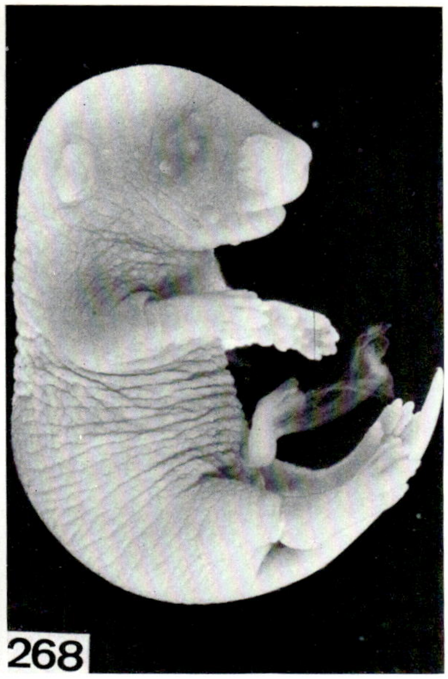

268

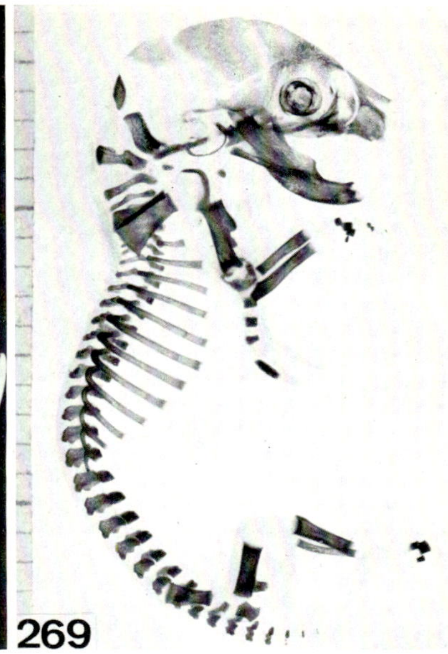

269

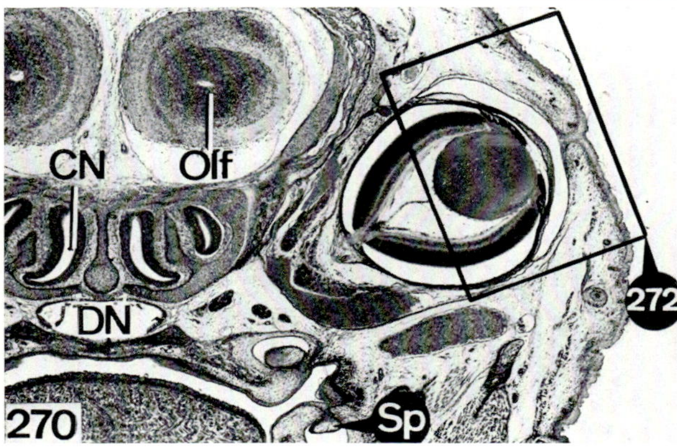

270

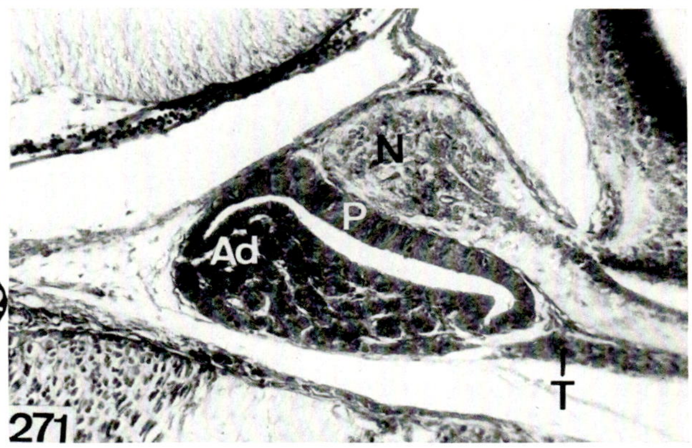

271

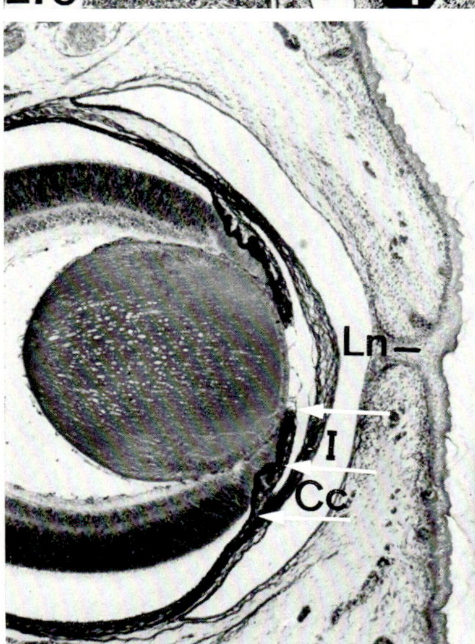

272

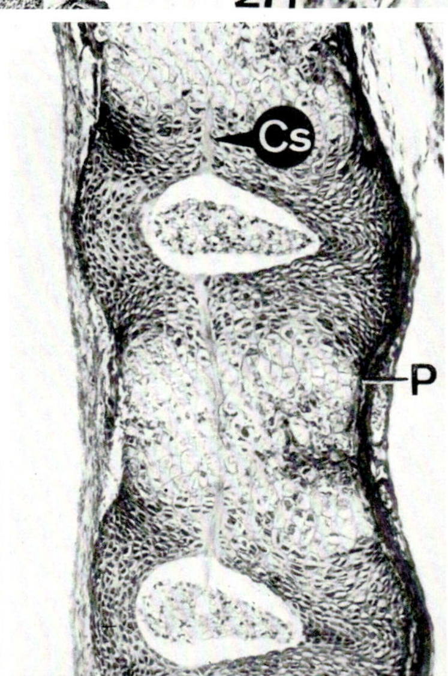

273

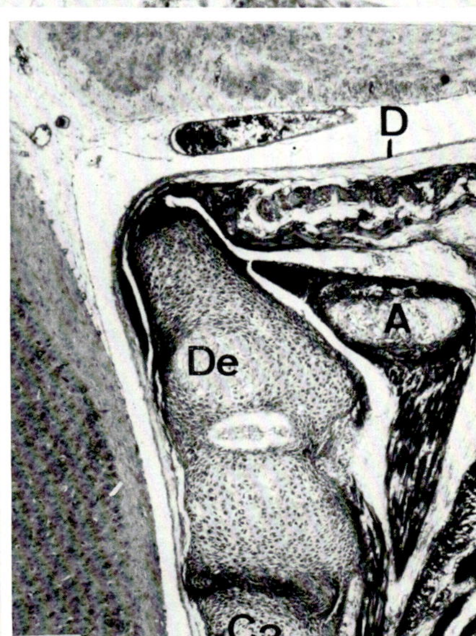

274

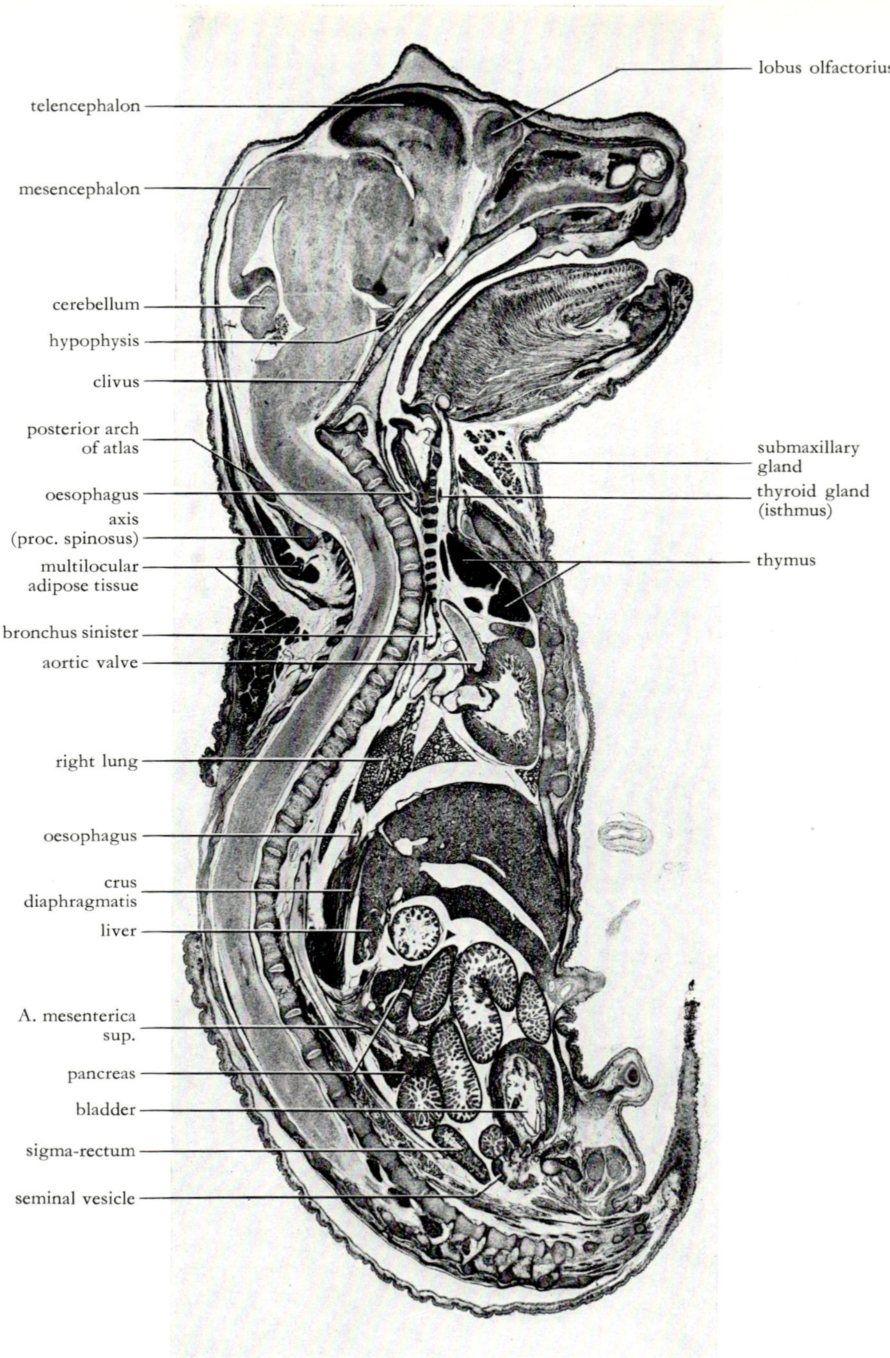

Fig. 275
Sagittal section, male fetus, 18 days 4 h, 23 mm.
KT 1041

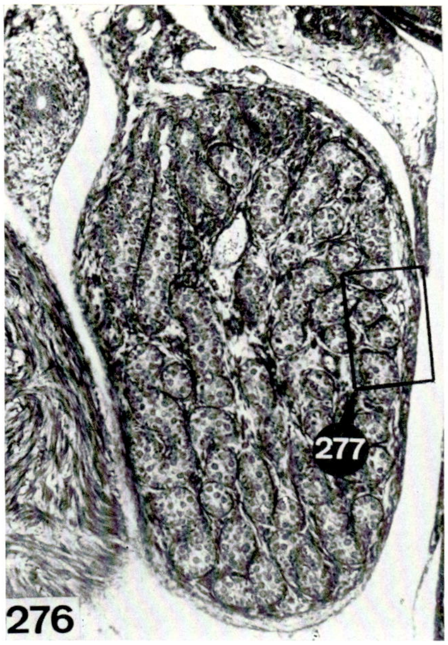

Fig. 276
Sagittal section through testis, 17 days 23 h.
KT 1010, 105:1

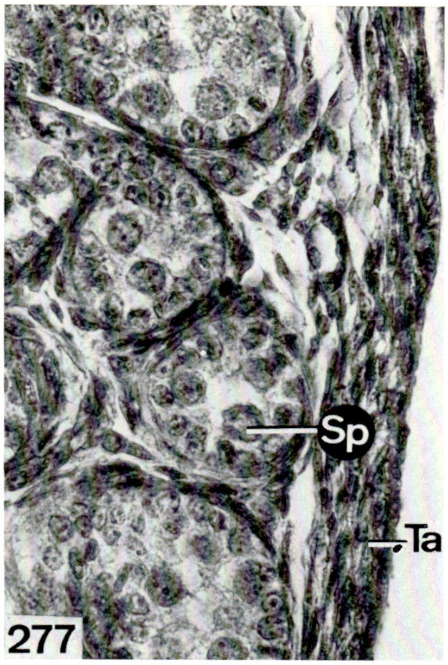

Fig. 277
High power view of testis Fig. 276.
Sp Spermatogonium, *Ta* tunica albuginea. 550:1

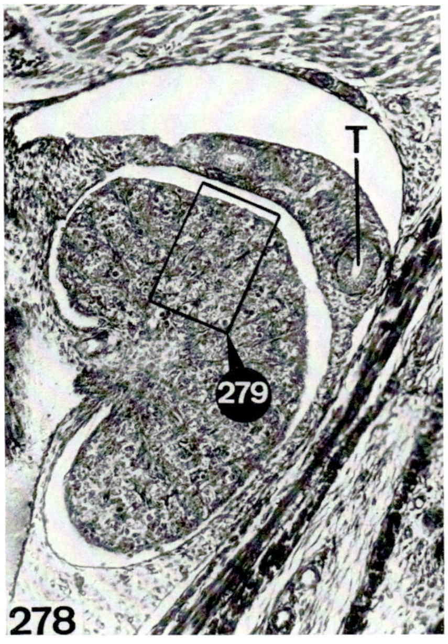

Fig. 278
Ovary, frontal section, 17 days 23 h.
T Oviduct KT 1010/30. 105:1

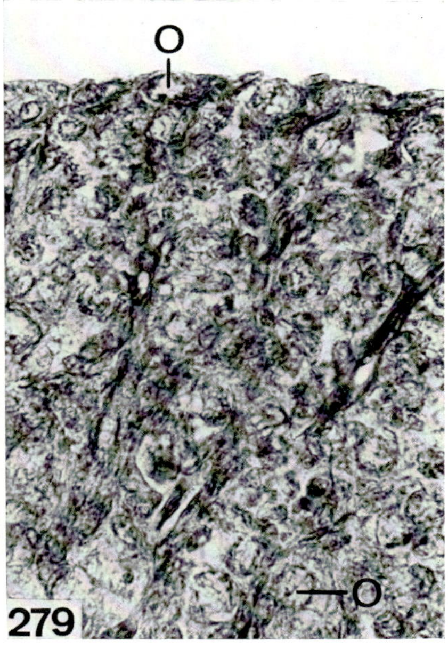

Fig. 279
High power view of ovary Fig. 278.
O Oocytes. 550:1

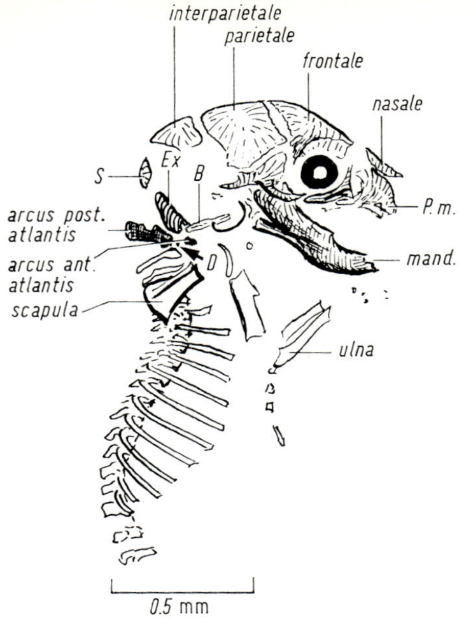

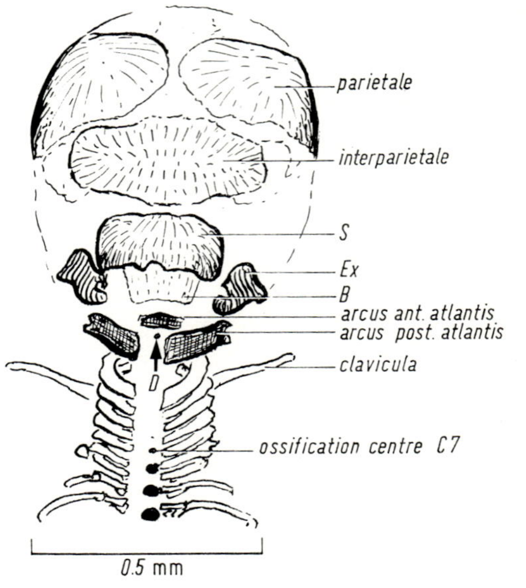

Figs. 280–281
Alizarin-cleared preparations, 18 days, 22 mm. Anterior part of the body in lateral (Fig. 280) and dorsal view (Fig. 281).
B Basioccipitale, *Ex* exoccipitale, *S* supraoccipitale, *D* (*arrow*) ossification center in basis of dens, *P.m.* premaxilla.
KT 1011

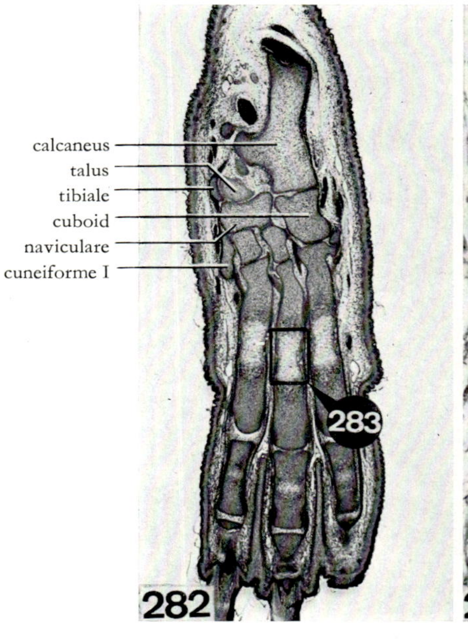

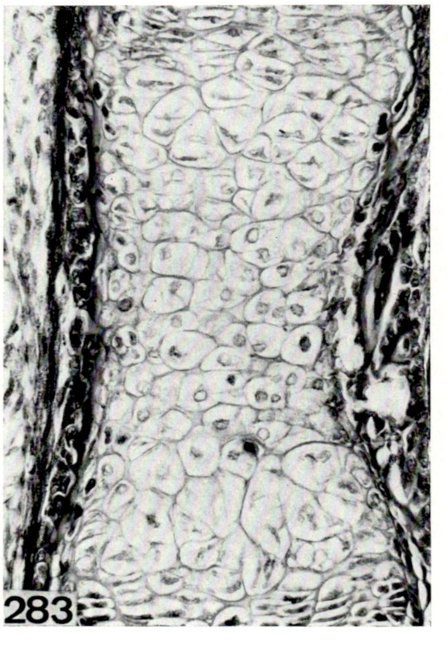

Fig. 282
Horizontal section through hindfoot plate, 17 days 23 h.
KT 1026. 22:1

Fig. 283
Enlarged view of metatarsale 3 (Fig. 282).
270:1

Stage 27 Newborn Mouse

19 Days, 23–27 mm.

External Features

The external appearance of the newborn mouse is not essentially different from the preceding stage (18 days), but the animal is considerably longer. The eyes and ears have closed. Histologically they are quite immature. The pigmented eye is still visible through the skin, which is thicker than in the 18-day fetus. The dark shadows of the liver and of the spleen (on the left side) can be seen through the body wall and the skin. Milk in the stomach can also be seen externally. The whiskers are long and clearly visible. The hair does not appear until the mouse is 2 or 3 days old.

The vagina remains closed for more than one month after birth.

Length: There is considerable variation in length, mainly due to different degrees of flexure of the body axis. Living newborns are usually 23 to 27 mm, but a few may be longer.

Circulatory System

The *ductus Botalli* (Fig. 293) has a tiny lumen for several hours after birth.

The blood-vessels of the lungs are considerably distended.

The foramen ovale (Fig. 294) is anatomically open, but functionally closed.

Near their entrance into the heart, the veins have distinct valves, i.e. at the termination of the vena jugularis externa, interna and at the inlet of the caval veins into the atrium (Fig. 294).

Intestinal Tract

The *oral cavity* has long been completely separated from the nasal cavities (Fig. 292).

The primordia of the incisors are shown in Fig. 292. The dental papilla (marked by the end of the line in Fig. 292) is bordered by tall odontoblasts. They already have formed some predentine. Peripherally, a layer of tall ameloblasts can be seen.

Derivatives of the Intestinal Tract

The *thyroid* (Fig. 295) is actively secreting. There are only two *parathyroids* in the mouse and they are bilaterally attached to the thyroid lobes. Displaced clusters of parathyroid cells are occasionally found within the connective tissue septa of the thymus [143].

The cortex of the *thymus* (Fig. 296) is subdivided into two zones:

1. an outer zone, containing large lymphoid cells, some of them in mitosis,
2. an inner zone, with many small lymphocytes.

In the medulla, there are clusters of large clear reticulum cells. Hassal's corpuscules are lacking.

Fig. 284	Newborn mouse, from the left, 27 mm. *L* Liver, *Sp* spleen, *M* stomach (filled with milk). 2.5:1
Fig. 285	Same specimen as in Fig. 284, right side, with millimeter scale. 2.5:1
Fig. 286	Alizarin-cleared preparation, newborn. *S* Supraoccipitale, *A* arcus posterior atlantis, *Sc* scapula dextra, *OJ* os ilii, *F* femur, *TC* talus and calcaneus. KT 891. 2.2:1
Fig. 287	First picture of dissection, newborn mouse. *T* Thymus; *H* heart, in pericardium; *Pc* lobus postcavalis of right lung; *L* liver; *B* bladder, filled. 2.5:1
Fig. 288	Second picture of dissection, same newborn mouse. *Sm* Glandula submaxillaris, *T* thymus, *P* pancreas, *M* stomach, *Sp* spleen, *Si* colon sigmoideum, *N* left kidney, *C* coecum, *J* ileum. 2.5:1
Fig. 289	Sagittal section through 4th and 5th thoracic vertebra, 19 days. *Cs* Notochordal sheath, *D* dura mater. 90:1
Fig. 290	Eye, frontal section, newborn mouse. *O* Nervus opticus, *Ln* suture of eye lids, *G* ganglion cell layer. KT 1067. 55:1
Fig. 291	High power view of iris (Fig. 290). *I* Iris, *Cc* corpus ciliare, *Mi* mitosis in lens epithelium. 270:1

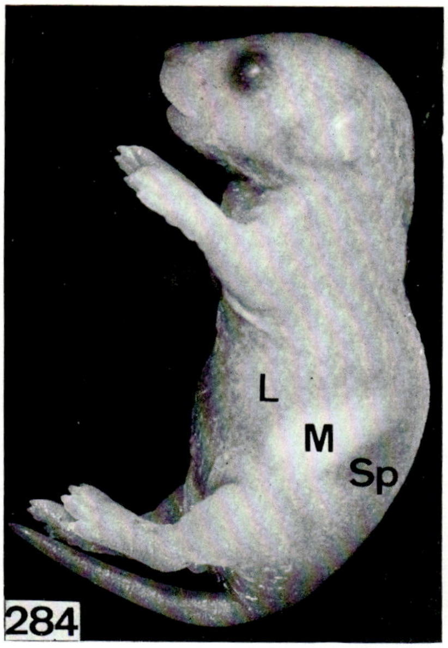
L
M
Sp
284

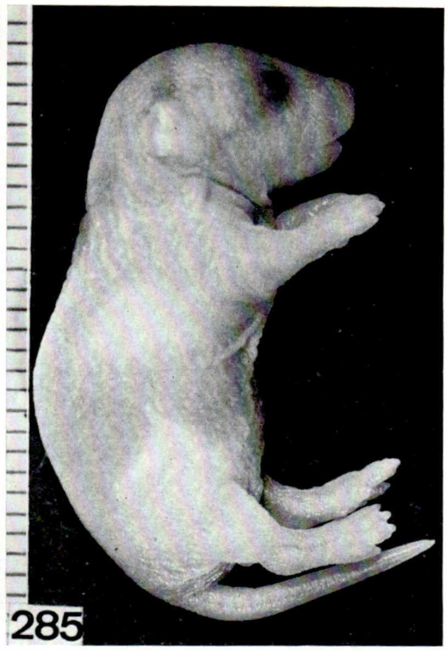
285

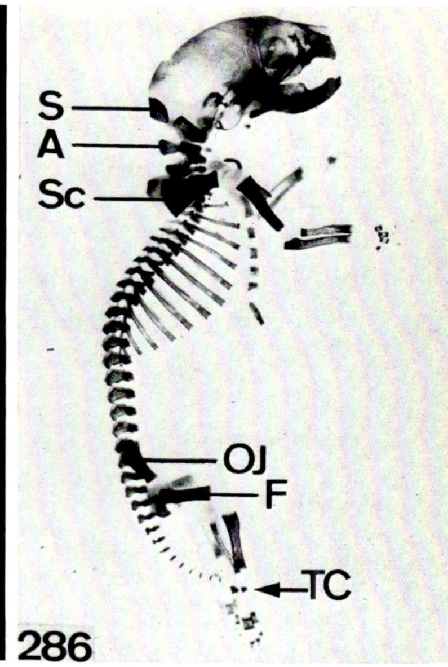
S
A
Sc
OJ
F
TC
286

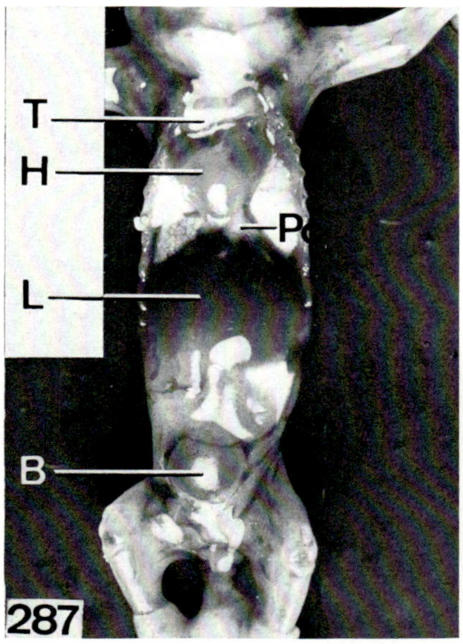
T
H
Pc
L
B
287

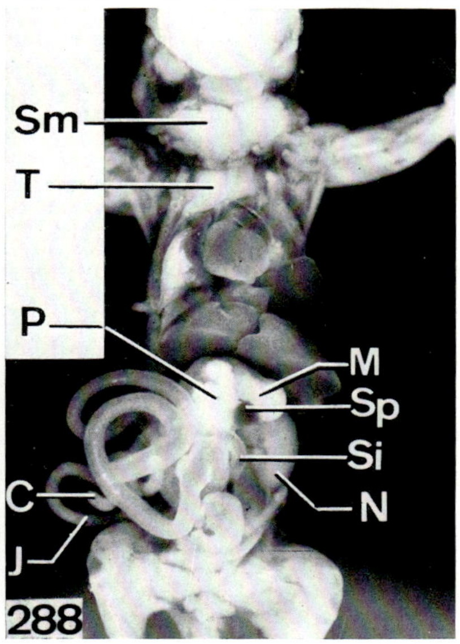
Sm
T
P
M
Sp
Si
C
N
J
288

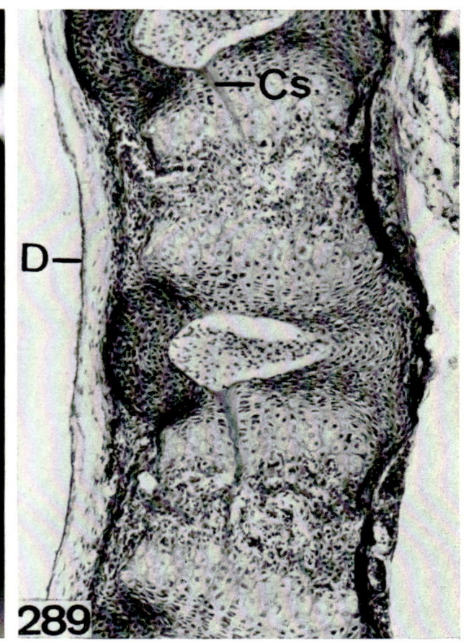
Cs
D
289

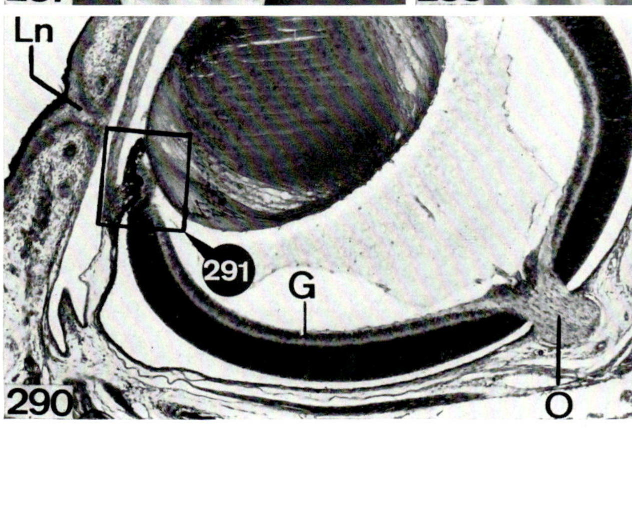
Ln
291
G
O
290

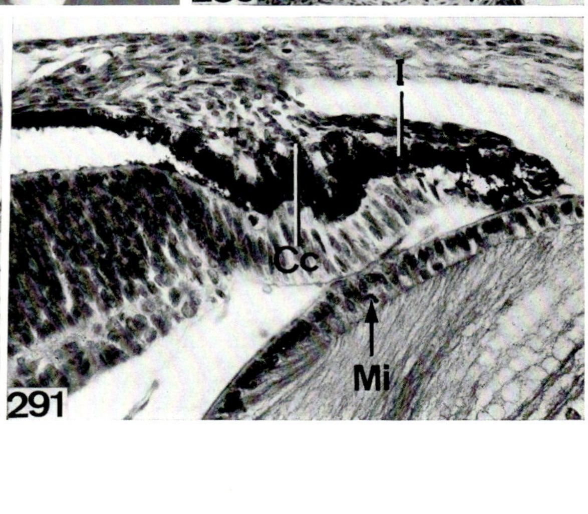
I
Cc
Mi
291

Respiratory System

The left *lung* consists of one undivided lobe. The right lung is subdivided into 4 lobes. In addition to the upper-, middle- and lower lobes there is a special lobus postcavalis. It is behind the heart, and extends quite far into the left side of the body (Fig. 294). Each bronchus lobaris gives off regular segmental and subsegmental bronchi (Fig. 301). The latter are connected to short terminal bronchi which open into wide alveolar sacs. There is a gradual transition between the cuboidal epithelium of the terminal bronchi and the flattened cells lining the alveoli. Except for the stem bronchi, the bronchial walls of the mouse do not contain cartilage.

Abdominal Viscera

The *liver* and *spleen* are sites of active hemopoiesis. The most conspicuous type of hemopoietic cells is the megakaryocyte and it is found in both organs. In the spleen, a few nodules of lymphatic tissue have formed. These are more distinct after birth [57].

The *stomach* is relatively large. It shows signs of marked secretory activity. The duodenum is the widest part of the intestine, and it has numerous villi. The large intestine is of the same calibre as the small intestine (Fig. 292).

The *adrenal glands* (Fig. 292) are well developed. The cortex and medulla are now more distinctly demarcated than at 17 days (Fig. 298). The medullary cells are smaller than the cortical cells.

In the cortex, three zones may be distinguished (Fig. 298):

1. A narrow outer zone, composed of small, crowded cells.
2. A large middle zone, consisting of large cells with abundant cytoplasm in which vacuoles (dissolved lipids) can be seen sometimes. The cells are not arranged in parallel strands.
3. A narrow inner zone with small, intensively stained cells. They are the precursors of the X-zone, which will develop after birth. It will regress at the age of one month.

In the medulla, the peripheral cells (bordering the cortex) are often clustered together reflecting their original arrangement in small groups. Between them there are intensively stained strands of cortical cells. The clusters of young medullary cells should not be confused with hemopoietic foci. Their nuclei are larger and lighter than erythroblastic nuclei, and their cytoplasm is basophilic. Furthermore, erythropoietic foci never occur in the cortex. Clusters of migrating young medullary cells may be encountered in the fetal cortex.

Urogenital Tract

There is a wide peripherally situated growth zone in the kidneys.

Diagnosis of sex: The perineum is definitely longer in the male. The distance between anus and genital papilla in the female is roughly one half of that in the male.

The topography of the *internal genital organs* is shown in simplified form in sketches of reconstructions (Figs. 303 and 304). The large seminal vesicle opens into a common orifice with the ampulla ductus deferentis into the urethra, at the colliculus seminalis. Later, a separate opening is often observed. The coagulating gland can be seen as a short sprout, situated immediately anterior (proximal) to the colliculus seminalis (Fig. 303). Other sprouts visible in Fig. 303 are primordia of the prostate (not labeled).

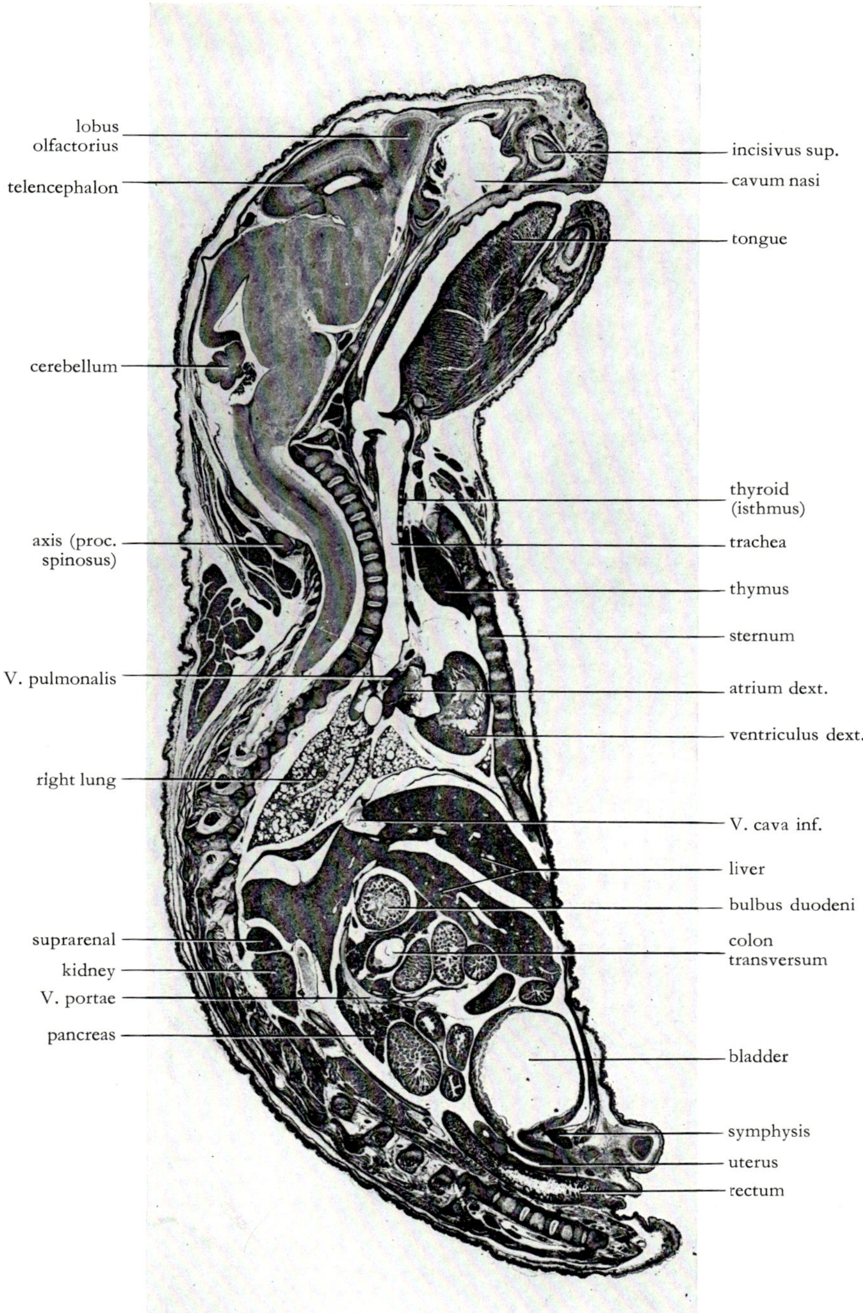

Fig. 292
Sagittal section through newborn female mouse.
KT 1067

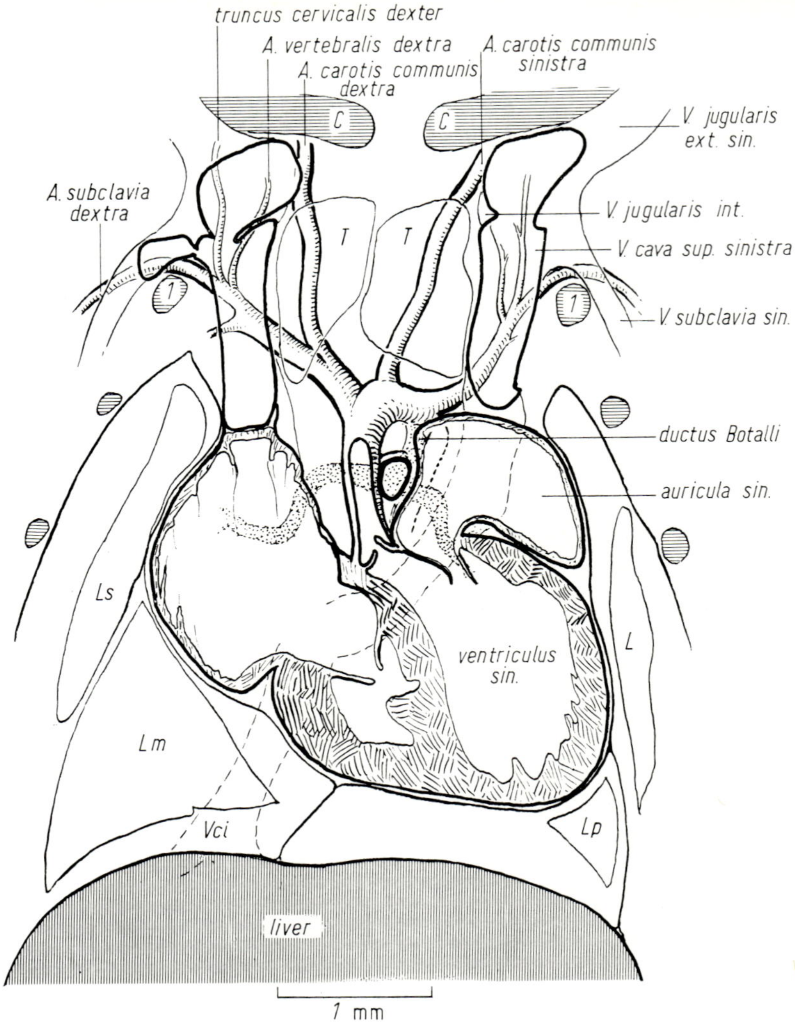

Fig. 293
Reconstruction of aorta and pulmonalis, superimposed on frontal section, ventral view. Pulmonalis and ductus Botalli stippled, contours of large veins are indicated. *C* Clavicula, *1* costa 1, *T* thymus, *L* left lung, *Lp* lobus superior postcavalis (pulmonis dextri), *Lm* lobus medius, *Ls* lobus superior, *Vci* vena cava inferior (projection).
KT 1067, newborn

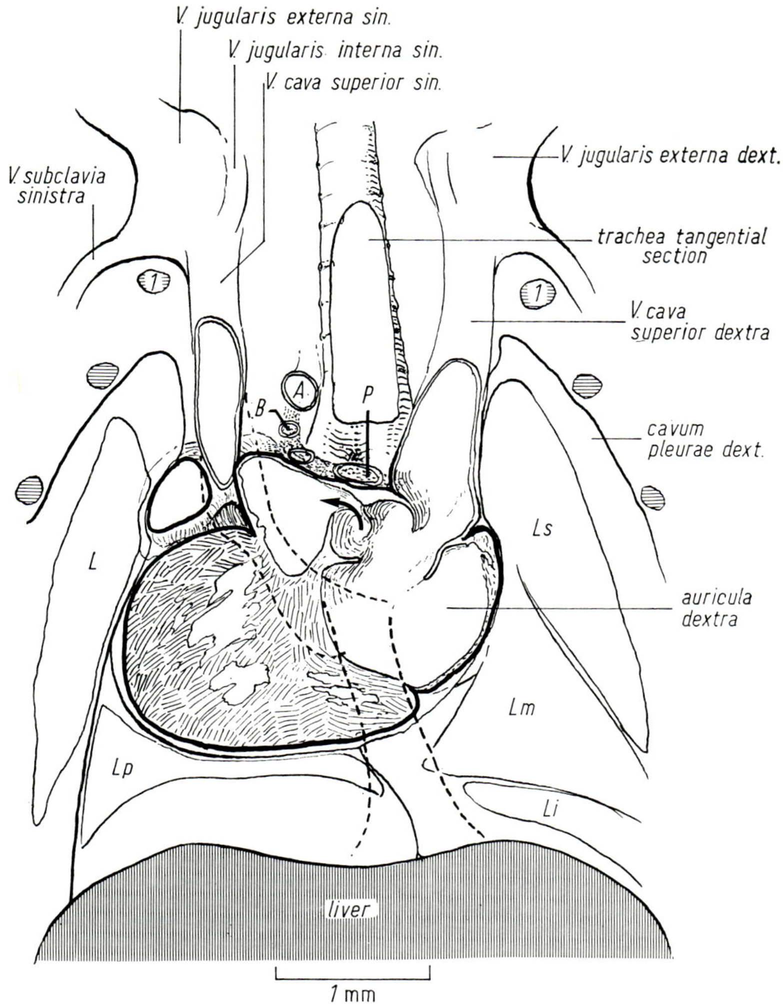

Fig. 294
Reconstruction of venous inlet, superimposed on frontal section. Dorsal view. Broken lines: openings of caval veins. *A* Arcus aortae, *P* arteria pulmonalis dextra, *B* ductus Botalli, *L* left lung, *Ls* lobus superior, *Lm* lobus medius, *Li* lobus inferior, *Lp* lobus postcavalis.
KT 1067

Fig. 295 Thyroid, isthmus, newborn mouse.
F Follicle. 550:1

Fig. 296 Thymus, newborn mouse.
Arrow: mitosis of a lymphoblast. 550:1

Fig. 297 Liver, newborn mouse.
Me Megakaryocyte, *P* branch of portal vein. 270:1

Fig. 298 Suprarenals, low power view, newborn mouse. 220:1

Fig. 299 High power view of boundary zone cortex–medulla (Fig. 298). 550:1

Fig. 300 Pancreas, newborn mouse.
In Pancreatic island.
KT 910. 550:1

Fig. 301 Frontal section through left lung, newborn mouse.
KT 1067. 40:1

Fig. 302 High power view of lung subsegment.
Sa Saccus alveolaris. 270:1

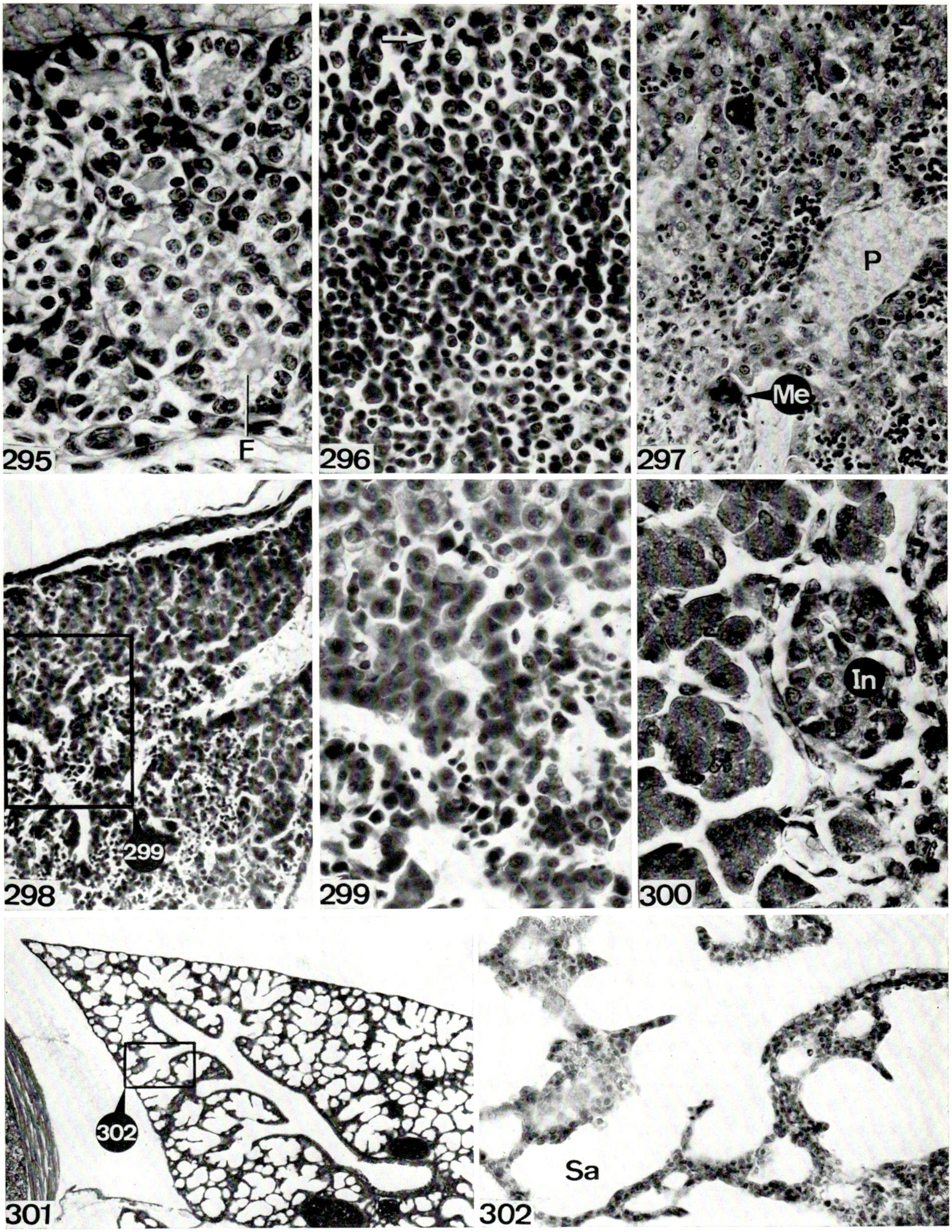
295
F
296
297
P
Me
298
299
299
300
In
301
302
302
Sa

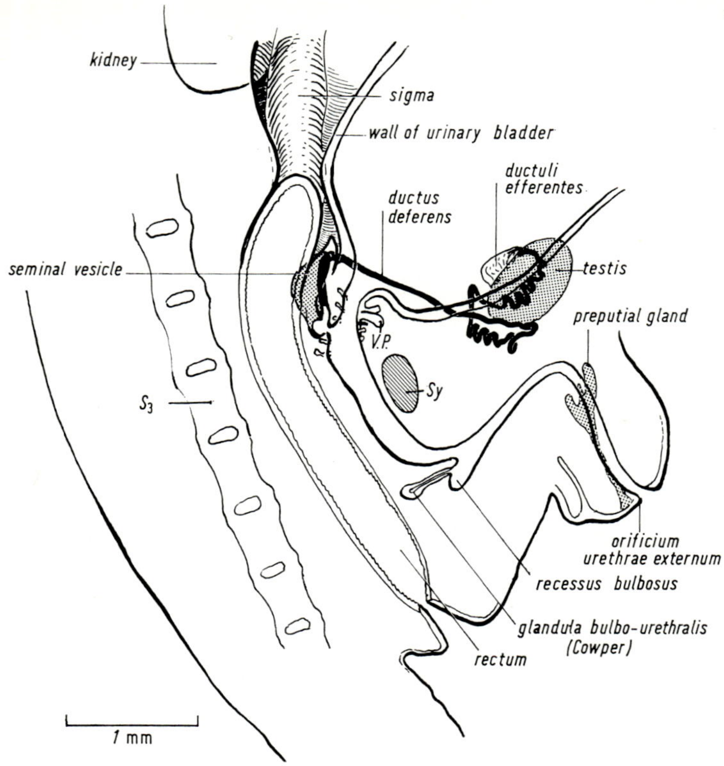

◀ Fig. 303
Male genital organs. Reconstruction, newborn mouse. Left half, viewed from the right. Caput and cauda epididymidis shown in simplified form. *V.P.* Ventral prostata, *Sy* symphysis, S_3 3rd sacral vertebra.
KT 910

Fig. 304
Female genital organs. Reconstruction, newborn mouse. Left half, viewed from the right. Oviduct shown in simplified form. *Sy* Symphysis.
KT 1067
▼

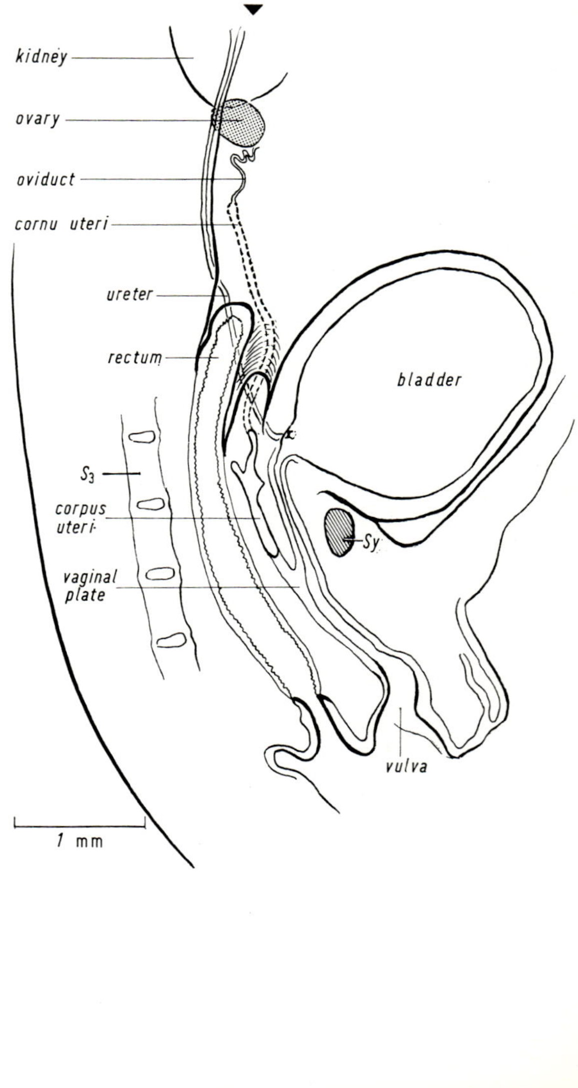

In the *ovary* many cells with condensed chromosomes can be seen, mostly prophases of the first meiotic division of the oocytes. Some primary follicles have developed in the more central area of the ovary. In this region, the oocytes are in the dictyotene stage. Graafian follicles are absent. The surface of the ovary is exposed to the fluid-filled space of the bursa ovarica, which is almost completely closed [107].

In the *testis* there are many interstitial cells. The tubuli contorti are composed of numerous primitive Sertoli cells (nourishing and supporting cells) and large gonocytes (Fig. 305). The *gonocytes* are no longer dividing. Mitotic activity will be resumed during the first week after birth. Simultaneously, the gonocytes will migrate towards the basement membrane of the seminiferous tubules. The developing daughter-cells will be

Fig. 305
Section through testis of newborn mouse. *T* Tunica albuginea, *G* gonocyte, *Z* interstitial cells. 550:1

Fig. 306
Section through ovary of newborn mouse. *O* Oocyte within primary follicle, *Mi* mitosis (prophase of first maturation division), *K* epithelium of ovary. KT 1067. 550:1

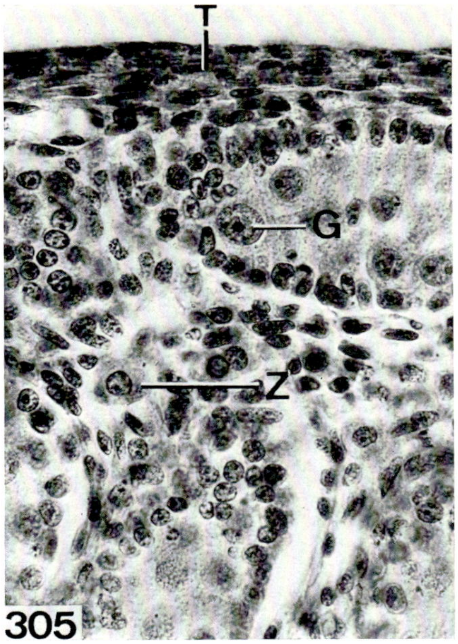

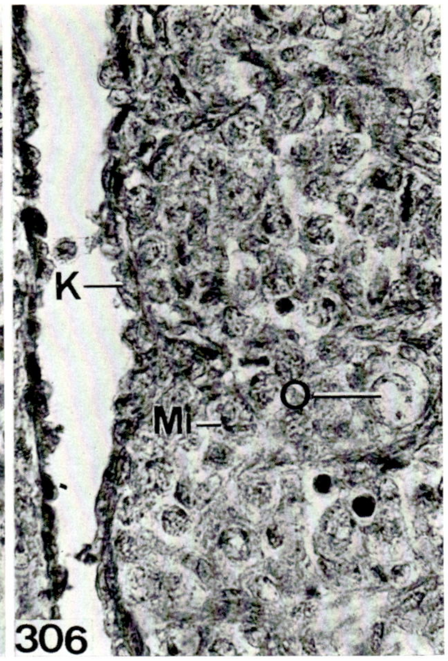

smaller and are called immature spermatogonia type A. Mature spermatogonia will remain smaller. These cells are generally flattened and closely attached to the basement membrane. They appear at the inition of spermatogenesis.

Central Nervous System

In the *cerebellum*, the fissures which developed in the previous stage have deepened considerably (Fig. 292). The structure of the hypophysis [136], which can be used as an indication of developmental age, has not changed, compare Fig. 271 (18 days).
The accessory organs of the *eye* have appeared. The conductory system of the infraorbital gland is composed of numerous branches. The passage opens into the saccus conjunctivae, together with the neighboring solid sprouts of the lacrimal gland. The Harderian gland reaches the conjunctiva by a single, slender duct, in the neighborhood of the nictitating membrane. These glands are not differentiated histologically. The nictitating membrane, however, contains some cartilage.
The iris and corpus ciliare (Fig. 291) remain at the state of maturation attained one day earlier (18 days, Fig. 272).
The *internal ear* is only slightly changed since day 18. The external auditory meatus has closed. The auditory apparatus cannot function until the second week after birth, when the organ of Corti has completed its maturation [180].

Skeletal System

The *vertebral column* has changed between the stage of 18 days and the newborn but the skull remains about the same. All ossification centers of the vertebral bodies of the trunk have formed, including cervical 3–5 (compare Fig. 307 to Figs. 280–281). Ossification is well advanced in the thoracic region (Fig. 289).

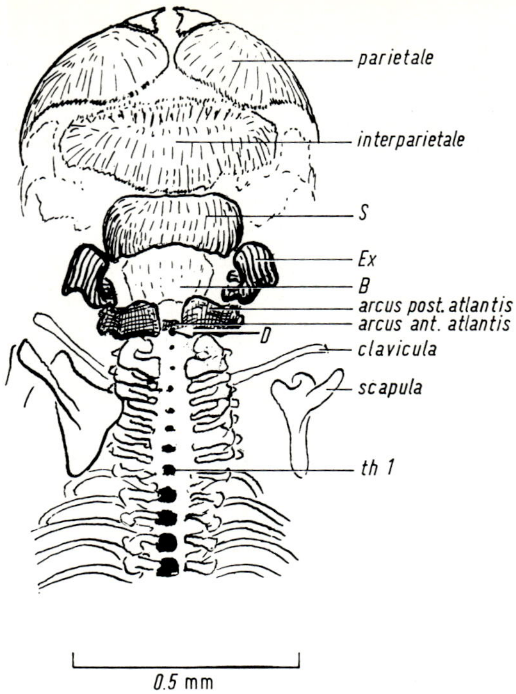

Fig. 307
Alizarin-cleared preparation. Anterior part of the body, dorsal view. Newborn.
S Supraoccipitale, *B* basioccipitale, *Ex* exoccipitale, *D* ossification center in basis of dens (*black point*).
KT 891

In the *sternum* the sternebrae have enlarged considerably. When the ribs grow slightly asymmetrical, the sternebrae are correspondingly asymmetrical ("crankshaft-sternum", Fig. 308). In the *fore-* and *hindfeet*, nearly all bones of the phalanges have formed. The metacarpalia and metatarsalia had already appeared previously. Only the centers of the fifth middle phalanx (hindfeet) and of the first phalanx (forefeet) are absent.
The *carpals* have not yet ossified. In the *tarsus*, however, 2 ossification centers can be seen regularly: the talus and calcaneus. They are distinctively large, even though they have just recently appeared (Fig. 286).

Placenta

The structure of the placenta has not changed fundamentally since the 14-day stage (Fig. 309). Reichert's membrane usually ruptures shortly before birth. When it ruptures, it often rolls up like a carpet. Where it borders the placenta, however, the membrane remains stretched (Fig. 311). In the vicinity of the placental margin, short processes in Reichert's membrane appear at 14 days (Fig. 309).

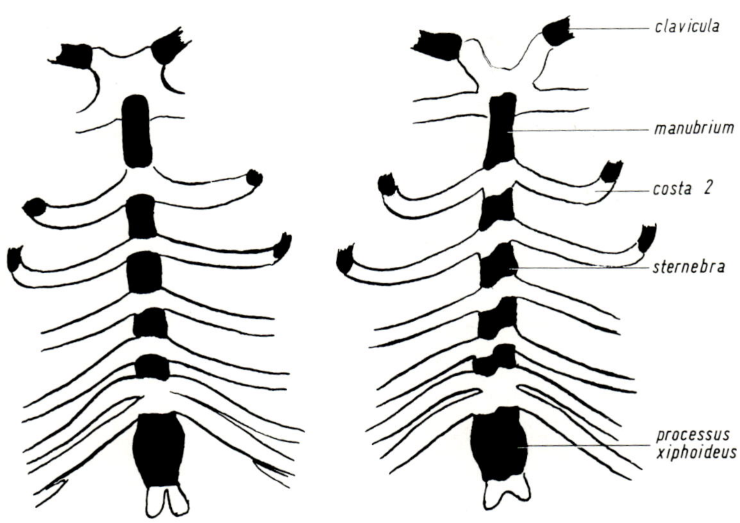

Fig. 308
Alizarin-cleared preparation of sternum. Note variable, but normal form of ossification centers ("crankshaft-sternum").
KT 865, 891

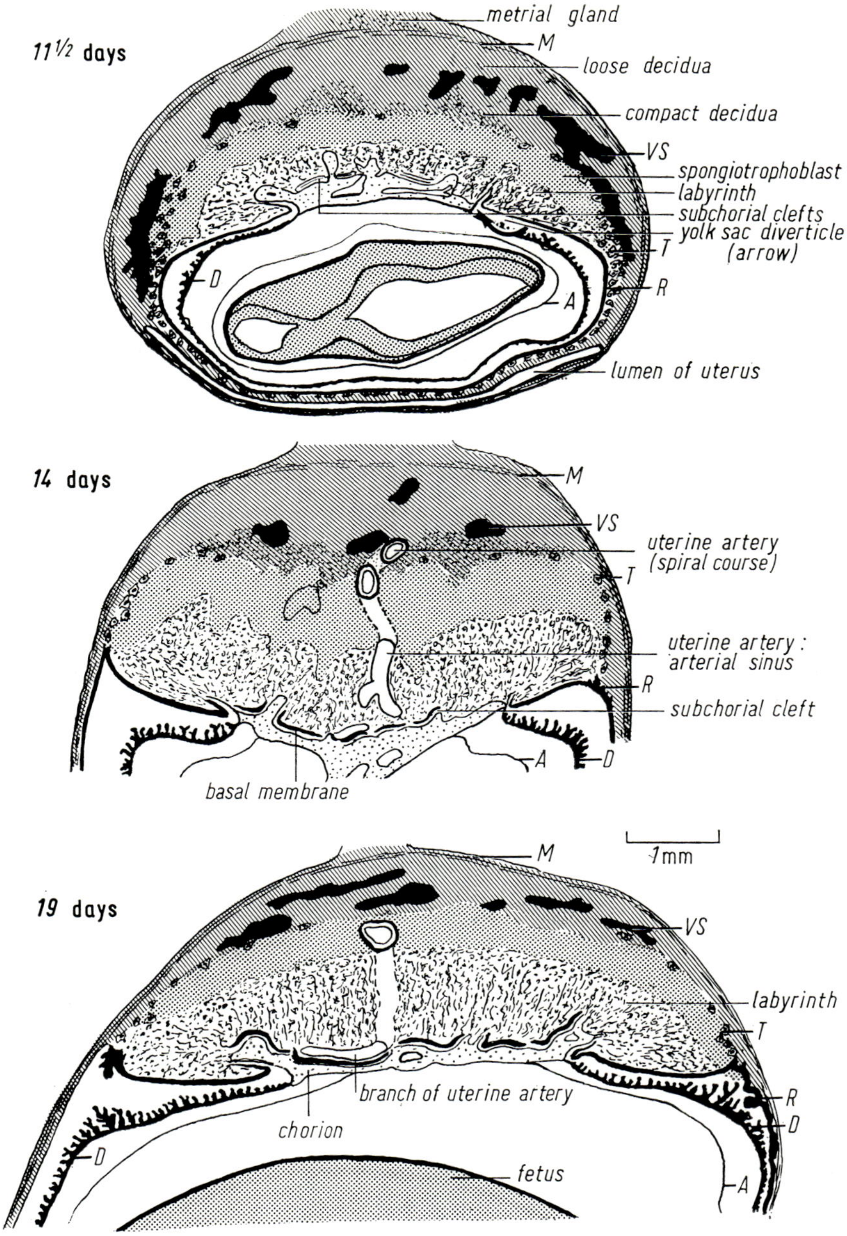

Fig. 309
Diagrammatic representation of placenta and extraembryonic membranes, from 11 1/2 days to birth.
M Muscle layer, *VS* venous sinus, *T* trophoblastic giant cells, *R* Reichert's membrane, *D* yolk sac, *A* amnion

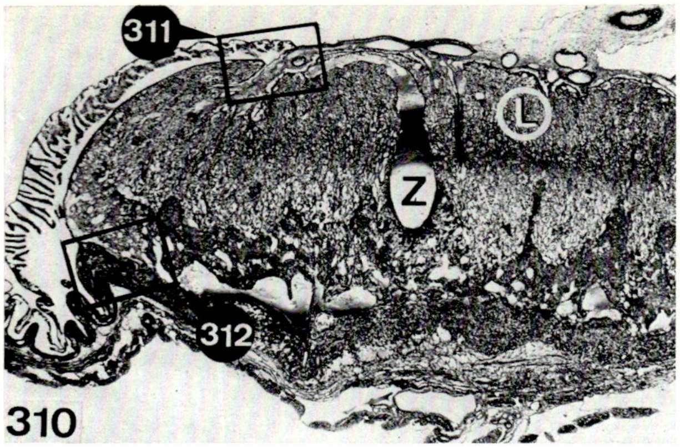

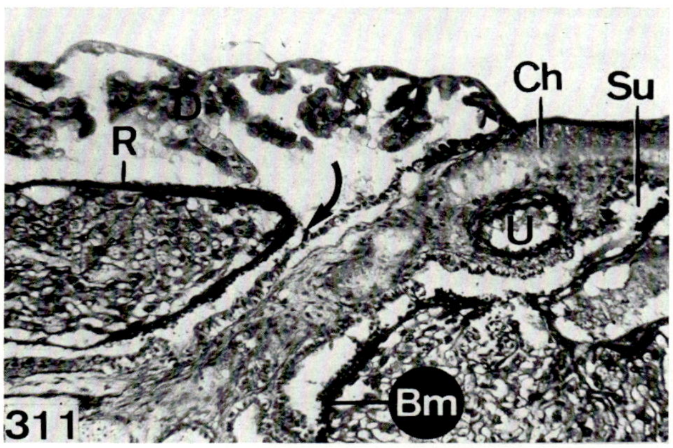

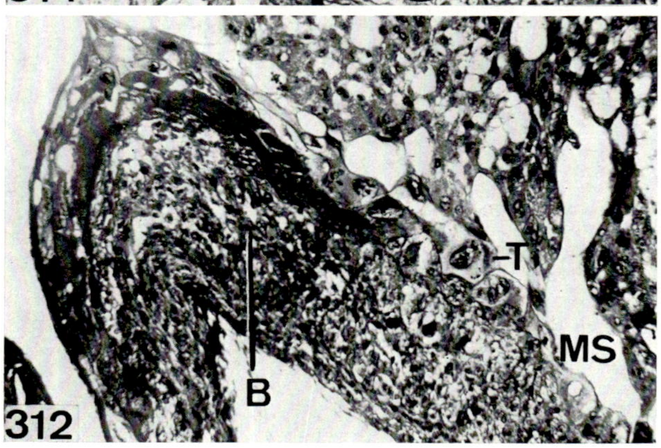

Fig. 310
Cross section through placenta of a newborn mouse, low power.
L Labyrinth, *Z* central artery.
KT 1056. 14.5:1

Fig. 311
High power view of chorionic plate (=*Ch*).
D Yolk sac, visceral layer; *R* Reichert's membrane; *Bm* basal membrane; *U* umbilical vessel; *Su* subchorionic space. *Arrow:* "yolk sac diverticle". 105:1

Fig. 312
High power view of basal plate (*B*).
MS Maternal blood sinus, *T* trophoblasticgiant cell. 105:1

At *19 days*, the placenta has not enlarged much. A dilated central artery enters the well developed labyrinth from the mesometrial side of the placenta. This vessel conducts maternal blood directly under the chorionic plate, where the stems of the umbilical vessels pass. The labyrinth is demarcated from the chorionic plate by the narrow intraplacental space [33]. This space consists of subchorionic clefts bordered by two walls of characteristic structure. Towards the chorion, one wall is lined by a layer of cuboidal cells. Towards the labyrinth, the other wall is lined by flattened epithelial cells supported by a thick basement membrane (Fig. 311). The intraplacental space opens laterally into the lumen of the yolk sac [33], ("yolksac diverticles", marked by *arrow* in Fig. 311).

The *labyrinth* is composed of an intricate maternal blood space and numerous fetal blood-vessels. Maternal and embryonic blood is separated by three cell layers, as shown by electron microscopic studies [34]. Two syncytial layers, which are closely approximated, are covered, near the maternal blood space, by single large trophoblastic cells.

The *spongiotrophoblast* (= reticular zone or junctional zone) [38] contains exclusively maternal blood-vessels. Large veins open from the decidua into extended (maternal) lacunae.

At the margins of the placenta, the spongiotrophoblast is bounded by the characteristic *giant cells* [39] mentioned above (Fig. 312).

Stage 28 Postnatal Development

Within 3 days, a fine fur appears, and the *ears* are opening. The internal ear is still quite immature. The definitive histologic structure does not develop until 13 days. The *retina* is also immature.

The *eyes* open at 12 to 14 days.

At 6 weeks, the vagina opens, and females begin estrous cycles. The first successful mating takes place at 2 or 3 months of age.

The *skeleton* is used here as a convenient index of relative stages of postnatal development. Mice of 7 and 24 days were chosen arbitrarily to be described.

7 Days Post Partum

The linear length (head to base of the tail) is about 38 mm.

At 7 days the distal epiphyseal centers appear in the tibia and fibula (Fig. 315).

In the tail, 24 *vertebrae* have ossification centers. Our adult hybrids have:

30–31 tail vertebrae
4 sacral vertebrae
6 lumbar vertebrae
13 thoracic vertebrae
7 cervical vertebrae

The neural arches of the sacral vertebrae are about to fuse, while the lumbar arches are still well separated (Fig. 314). The ossification centers of the vertebral bodies are separated throughout from the centers of the arches by narrow epiphyseal plates (Fig. 318). The discrete horizontal cleft separating the two centers of the axis (i.e. base of the body and base of the dens) is difficult to recognize (Fig. 314).

Extremities

Since birth, numerous ossification centers have appeared as seen in Figs. 315 and 317. The distal epiphyseal centers in the *tibia* and *fibula* are typical for 7-day mice. They appear shortly after the distal epiphyseal center of the femur.

In the tarsus, not only the talus and calcaneus are visible in alizarin-stained cleared preparations, but also the cuboid, 3 cuneiformia and naviculare. The os tibiale, however, has not yet ossified.

In the *ovary*, there are abundant primary follicles in the cortical zone. In deeper areas, some (secondary) follicles are growing, but Graafian follicles are still lacking (Fig. 316).

There is little change in the *testis* since birth. Many cells have entered prophase. Spermiogenesis will start at 9 days of age.

In the *thymus* a few small Hassal's corpuscules can be seen (Fig. 319).

Fig. 313 7-day old mouse, female.
KT 893. 1.6:1

Fig. 314 Alizarin-cleared preparation, 7 days post partum, dorsal view.
S Supraoccipitale, *Ax* 2 ossification centers of axis (axis body and basis of dens), *At* processus transversus atlantis, *U* ossification center in olecranon (ulna), L_2 body of second lumbar vertebra, *OJ* os ilii, *Is* os ischii.
KT 893. 1.6:1

Fig. 315 Same skeleton as in Fig. 314, lateral view.
Fe Ossification center in distal epiphysis of femur; *TF* ossification centers in distal epiphysis of tibia and fibula, having newly arisen; *Ta* tarsalia (7 ossification centers, partially hidden). 1.6:1

Fig. 316 Ovary, showing secondary follicles, located near central region, 7 days post partum.
KT 892. 105:1

Fig. 317 Thorax, ventral view, 7 days post partum. Alizarin-cleared preparation.
Initial calcification of rib cartilage (*R*). *Tm* Apophyseal center in tuberculum maius, *E* epiphyseal center in caput humeri, *Co* ossification center in coracoid process (below clavicula), *Ma* manubrium sterni, *Xi* xiphoid process.
KT 893. 5:1

Fig. 318 Lumbar vertebrae, ventral view, Alizarin-cleared preparation, 7 days post partum.
L_2 Body of 2nd lumbar vertebra, *Ar* arch-center of 2nd lumbar vertebra, *OJ* os ilii, *Is* os ischii, *P* os pubis.
KT 893. 5:1

313

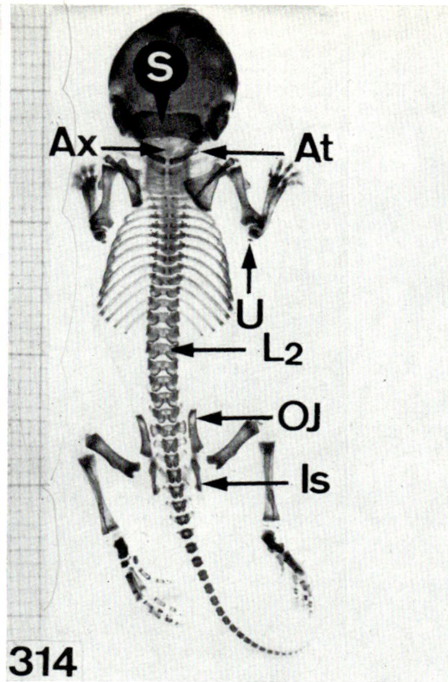

314

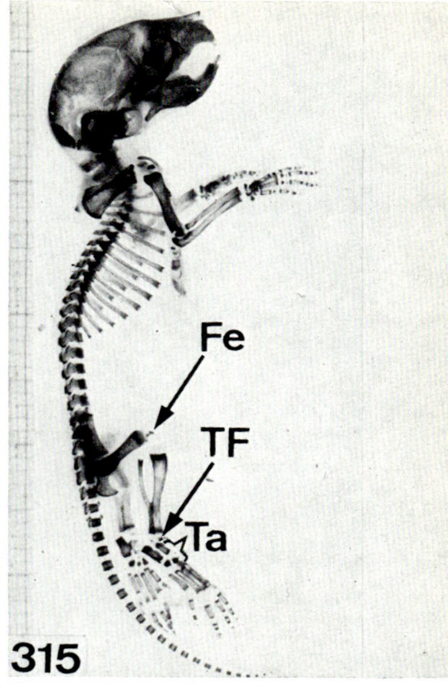

315

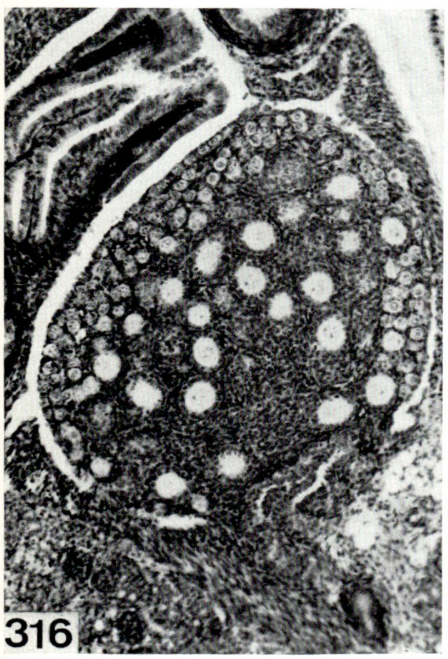

316

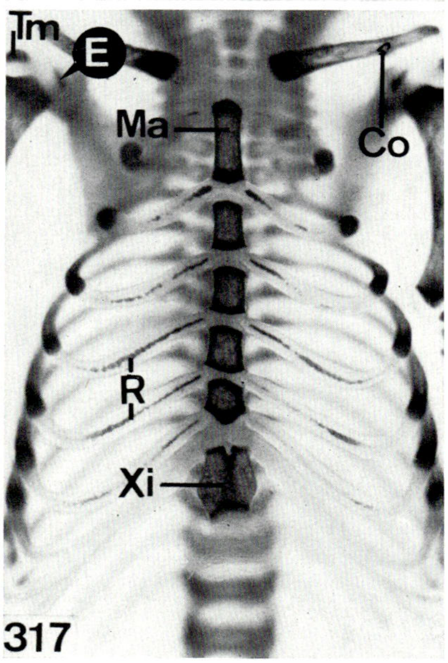

317

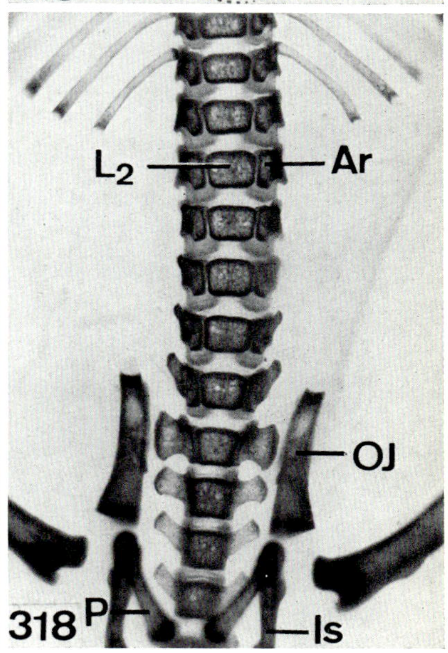

318

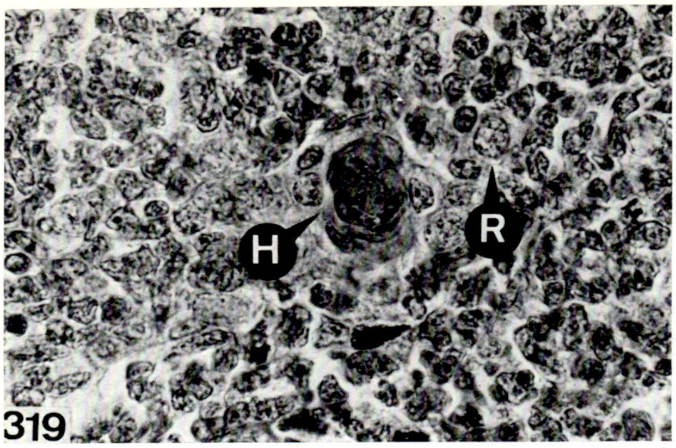

Fig. 319
Hassal's corpuscule (*H*) within thymus of 7-day old mouse.
R Reticulum cell.
KT 892. 700:1

In the *eye*, the external plexiform layer is beginning to form, and it can be recognized for the first time. As a consequence, the nuclear zone becomes subdivided into external and internal nuclear zone. The process of separation starts centrally, near the optic nerve, and proceeds quickly towards the periphery. On the 11th day after birth, the outer segments of the photoreceptor cells are forming.

At *14 days*, the rods have attained their final length.

In the internal *ear*, the organ of Corti is differentiating [180]. Differentiation starts at birth and is complete at 13 days (Figs. 324–326).

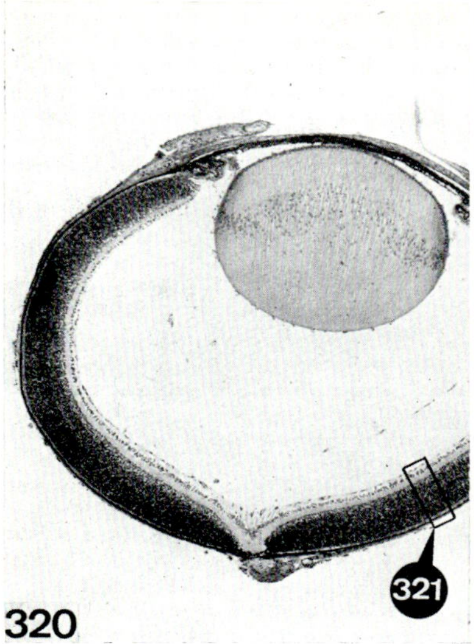

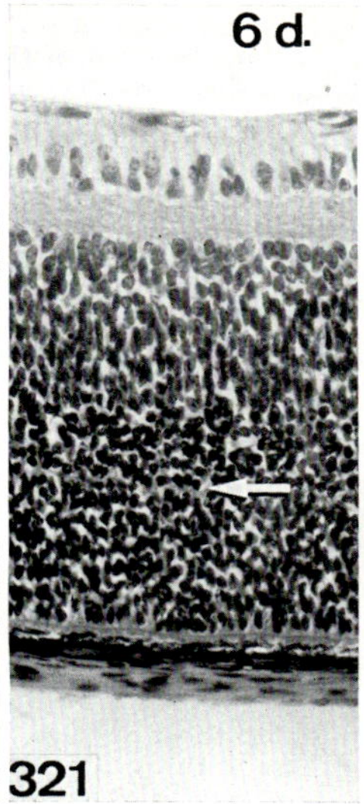

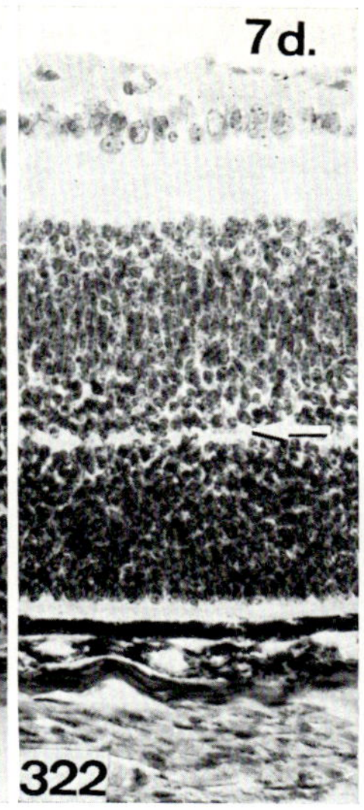

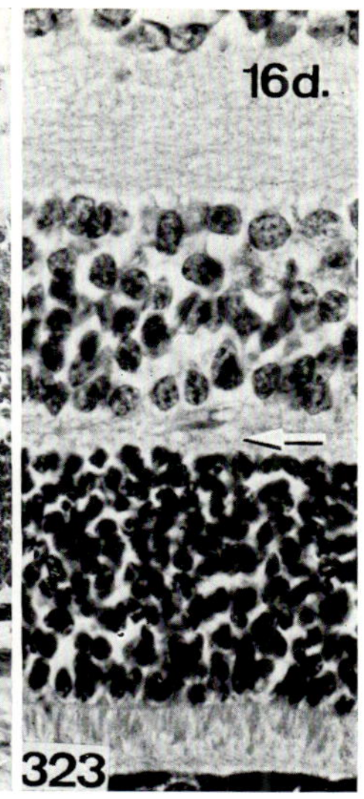

Fig. 320 Low power view of eye, 6 days post partum, horizontal section. 28:1

Fig. 321 High power view of retina of Fig. 320.
Arrows in Figs. 321–323: development of outer plexiform layer, separating inner and outer nuclear layers. 240:1

Fig. 322 At 7 days, the nuclear layers are distinctly separated.
KT 892. 240:1

Fig. 323 Fully differentiated retina, 16 days post partum. 500:1

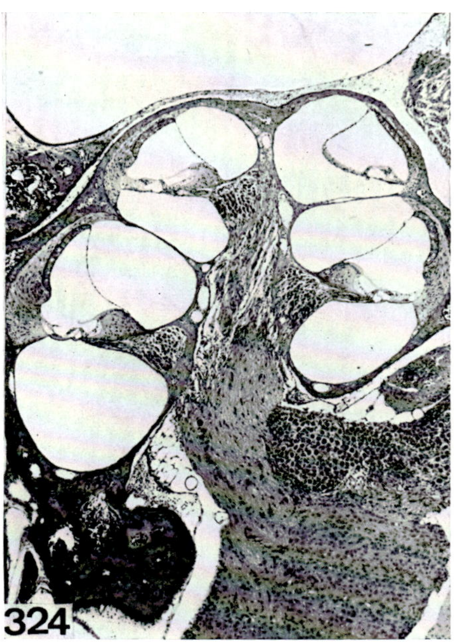

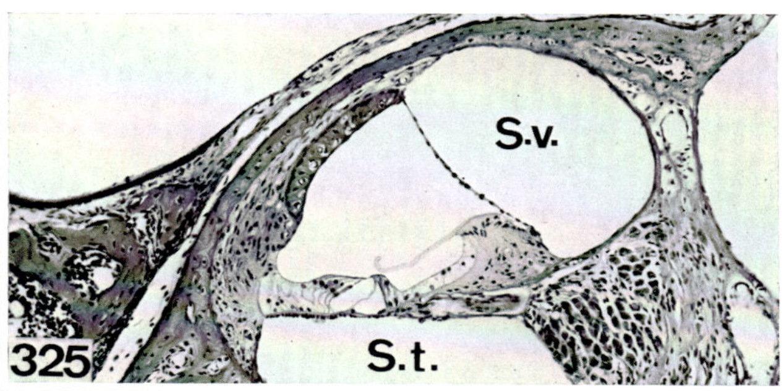

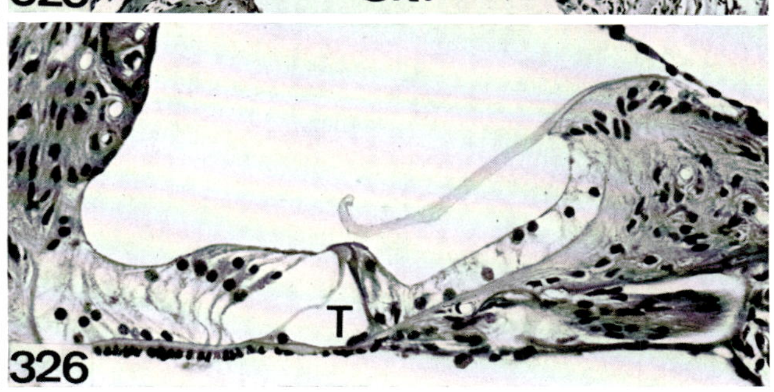

Fig. 324 Longitudinal section through axis of cochlea, 16 days post partum, low power view 40:1

Fig. 325 High power view of cochlear duct (apical part) of Fig. 324.
S.v. Scala vestibuli, *S.t.* scala tympani. 150:1

Fig. 326 High power view of organ of Corti of Fig. 325.
T Inner tunnel. 270:1

24 Days Post Partum

The linear length (head to base of tail) ranges from 50–60 mm.
At this age, the animals may be weaned.
The *vertebral column* continues to grow. In microscopic preparations, large cartilaginous growth zones are visible. The tail has attained its full number of 30 osseous vertebrae. There are only 4 pairs of hemal arch bones. These are small spherical pieces of bone bilaterally situated ventral to the caudal intervertebral discs. They develop first in the proximal part of the tail.
In the tarsus, the appearance of the apophysis of the tuber calcanei is a characteristic of this age (Fig. 329). In the lower and upper thigh of cleared animals, distinct epiphyseal plates can be seen. There are also epiphyseal plates in the acetabulum (Fig. 330).
Numerous blood vessels have grown into the flat epiphyses of the sacral and of the adjacent caudal vertebrae (Fig. 332).
In the *testes* there is active spermatogenesis. At 13 days after birth, differentiated *Sertoli cells* may be recognized. The original large and centrally placed primordial germ cells give rise to smaller, peripherally situated *spermatogonia.*
Many gonocytes have entered meiotic prophase. At the same time, the seminiferous tubules are developing lumina.
At 24 days there are no spermatozoa, but some tubules contain numerous young spermatids. Some peritubular cells are now transforming into smooth muscle cells (M. H. Ross [106]).
In the *ovary*, primary, secondary (= growing) and tertiary (= Graafian) follicles can be recognized (Fig. 331). The oocytes are in dictyotene stage. Occasionally, multinucleate eggs may be seen. I have seen them only in follicles containing numerous pycnotic and degenerating granulosa cells. Probably these follicles will soon degenerate. The growing follicle in Fig. 331 also shows signs of atresia.

Fig. 327 Hybrid female mouse, 24 days post partum

Fig. 328 Alizarin-cleared preparation, lateral view, male, 24 days post partum. Right extremities removed, millimeter scale.
KT 899

Fig. 329 Dorsal view of skeleton, before removing right extremities.
Ap Apophyseal center in tuber calcanei.
KT 899. 1.5:1

Fig. 330 Ventral view of skeleton, enlarged, 24 days post partum.
M Manubrium sterni, *H* acetabulum with epiphyseal plates.
KT 899. 2.7:1

Fig. 331 Section through ovary, 24 days post partum. Hematoxylin-Eosin.
T Tertiary follicle (Graafian follicle), *P* polynucleate oocyte, in secondary follicle.
KT 1071. 105:1

Fig. 332 Intervertebral disc of proximal tail. Frontal section at 4 weeks.
CH Notochordal sheath within epiphyseal plate of 2nd caudal vertebra, *G* blood vessels in epiphyseal plate of first caudal vertebra.
KT 787a. 105:1

327

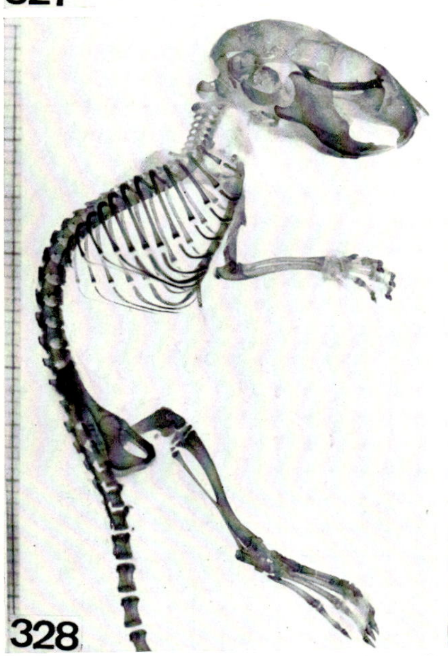

328

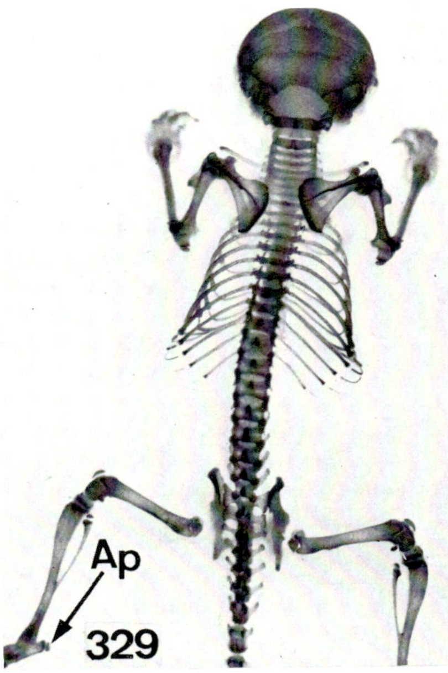

329

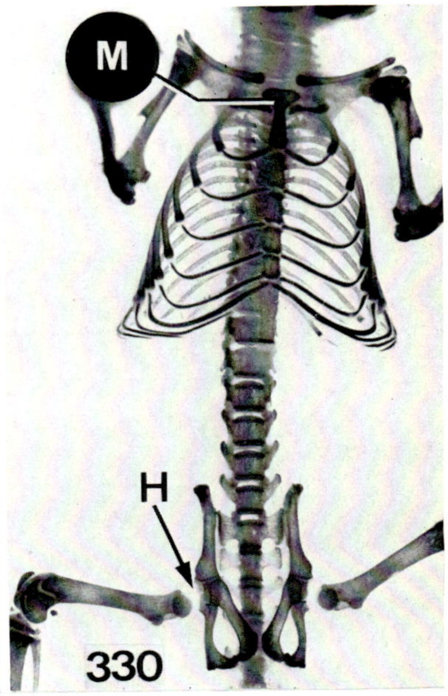

330

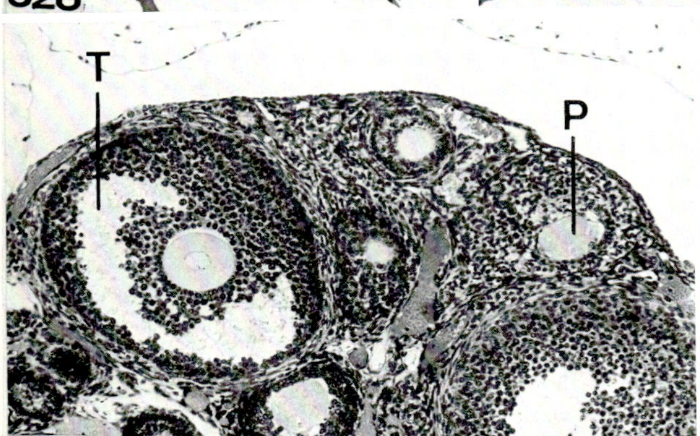

331

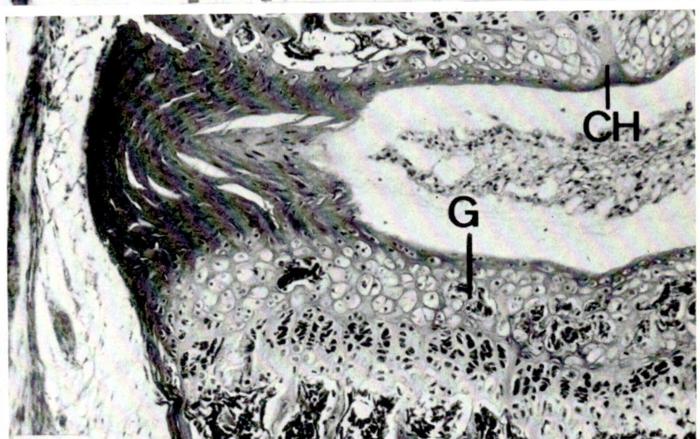

332

29 Weight Curves

The mouse has reached its full size at 3–4 months. The growth curve is dependent on genetic (Fig. 333) and environmental [3] factors, especially the amount of milk available [198]. The temperature also has some influence on growth [201].

The males become heavier than the females at the age of 4 weeks.

Specific pathogen-free (SPF) animals seem to grow faster than conventionally reared mice. Before we established our SPF colony, our CBA's had a considerably lower growth rate than the C57BL/6J animals. Now, in pathogen-free colonies, the CBA's grow faster than the C57BL/6J mice. Environmental factors may play an unexpected role in the growth rate of mice. They are difficult to take into account in applying a mathematical model [196] to the growth of inbred mice (Fig. 334).

Pregnant females begin to gain appreciable weight from the 8th day of pregnancy (Fig. 335).

Like growth rate, *life span* is influenced by many genetic and environmental factors. It may vary between 1 and 3 years [2].

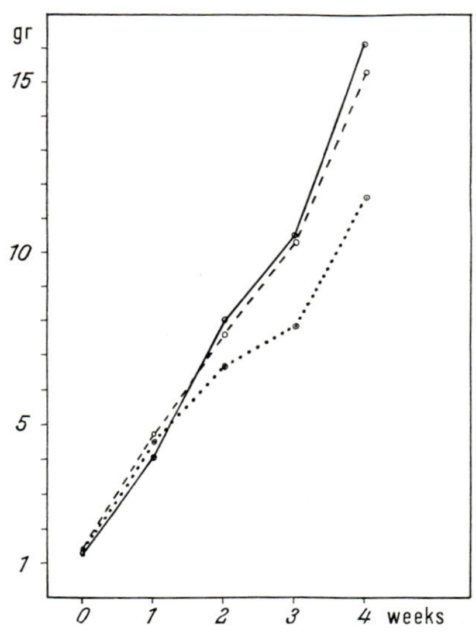

Fig. 333
Mean weights of our growing inbred and hybrid mice.
Hybrids (full line) show maximum increase in weight. Broken line: CBA. Stippled: C57BL/6

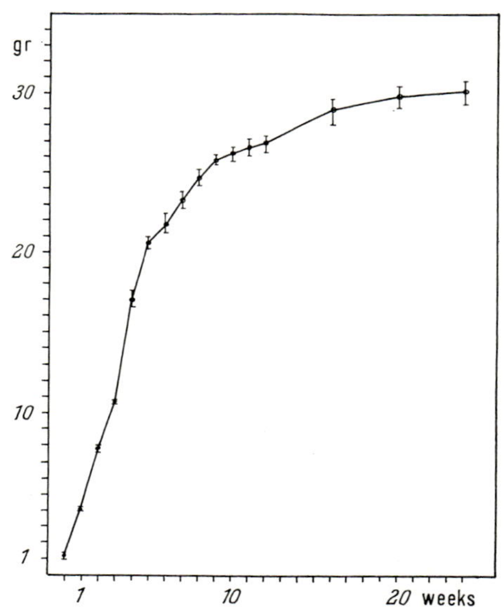

Fig. 334
Postnatal increase in weight of our hybrid males.
Short perpendicular lines: standard errors

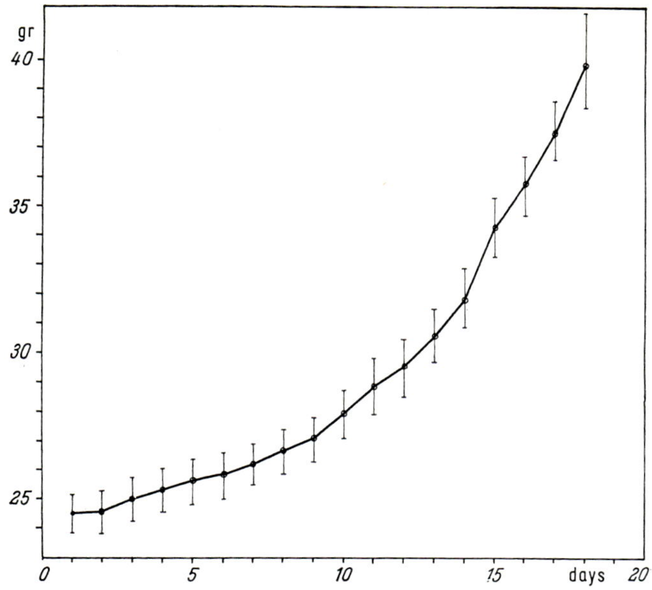

Fig. 335
Increase in weight of pregnant hybrids (C57 BL/6-females). Short perpendicular lines: standard errors

Selected References

Opera Generalia

1. Cook, M. J.: The anatomy of the laboratory mouse, 143 p. London-New York: Academic Press 1965.
2. Green, E. L.: Biology of the laboratory mouse, 706 p. New York: McGraw Hill 1966.
3. Gruneberg, H.: The genetics of the mouse, 2nd edit., 650 p. The Hague: Martinus Nijhoff 1952.
4. Otis, E. M., Brent, R.: Equivalent ages in mouse and human embryos. Anat. Rec. **120**, 33–64 (1952).
5. Rugh, R.: The mouse. Its reproduction and development, 430 p. Minneapolis: Burgess 1968.
6. Streeter, G. L.: Developmental horizons in human embryos. Contr. Embryol. Carneg. Instn. **30**, 213–230 (1942).

Preimplantation Period

7. Austin, C. R.: Fertilization, 145 p. Englewood, N.Y.: Prentice-Hall 1965.
8. Beatty, R. A., Sharma, K. N.: Genetic of gametes. III. Strain differences in spermatozoa from eight inbred strains of mice. Proc. roy. Soc. Edinb. B **68**, 27–53 (1960).
9. Beatty, R. A., Fischberg, M.: Heteroploidy in mammals. III. Induction of tetraploidy in pre-implantation mouse eggs. J. Genet. **50**, 471–479 (1952).
10. Calarco, P. G., Brown, E. H.: An ultrastructural and cytological study of pre-implantation development of the mouse. J. exp. Zool. **171**, 253–284 (1969).
11. Chang, M. C., Hunt, D. M.: Morphological changes of sperm head in the ooplasm of mouse, rat, hamster and rabbit. Anat. Rec. **142**, 417–426 (1962).
12. Dickson, A. D.: The form of the mouse blastocyst. J. Anat. (Lond.) **100**, 335–348 (1966).
13. Dickson, A. D.: Variation in development of mouse blastocysts. J. Anat. (Lond.) **101**, 263–267 (1967).
14. Edwards, R. G.: Timing of stages of maturation divisions, ovulations, fertilization and the first cleavage of eggs of adult mice treated with gonadotropins. J. Endocr. **18**, 272–304 (1959).
15. Graham, C. F.: Parthenogenetic mouse blastocysts. Nature (Lond.) **226**, 165–167 (1970).
16. Lewis, W. H., Wright, E. S.: On the early development of the mouse egg. Contr. Embryol. Carneg. Instn. **25**, 113–143 (1935).
17. Potts, D. M., Wilson, J. B.: The pre-implantation conceptus of the mouse at 90 hours post coitum. J. Anat. (Lond.) **102**, 1–11 (1967).
18. Stevens, L. C.: The development of teratomas from intra-testicular grafts of 2-cell mouse eggs. Anat. Rec. **157**, 328 (abstr.) (1967).
19. Weitlauf, H. M.: Temporal changes in protein synthesis by mouse blastocysts transferred to ovariectomized recipients. J. exp. Zool. **171**, 481–486 (1969).
20. Whitten, W. K., Biggers, J. D.: Complete development in vitro of the pre-implantation stages of the mouse in a simple chemically defined medium. J. Reprod. Fertil. **17**, 399–401 (1958).
21. Wolstenholme, G. E. W., O'Connor, M.: Pre-implantation stages of pregnancy. Ciba Foundation Symposium, 430 p. London: J. & A. Churchill Ltd. 1965.

Implantation and Fetal Membranes

22. Amoroso, E. C.: Placentation. In: A. S. Parkes (edit.), Marshall's physiology of reproduction, 3rd edit., vol. 2, p. 127–311. London: Longmans, Green 1952.
23. Arvis, G.: Les cellules ultratrophoblastiques. Etude ultrastructurale au cours de la nidation chez la souris. C. R. Ass. Anat. 55e Congr. No 147, 87–104 (1970).

24. Billingham, W. D., Kirby, D. R. S., Owen, J. J. T., Ritter, M. A., Burtonshaw, M. D., Evans, E. P., Ford, C. E., Gauld, I. K., McLaren, A.: Placental barrier to maternal cells. Nature (Lond.) **244**, 704–706 (1969).
25. Bloch, S.: Die Glandula myometralis im Uterus der Maus. Acta anat. (Basel) **56**, 103–119 (1964).
26. Bloch, S.: Beobachtungen zur Wechselwirkung zwischen Keim und Uterus bei der Implantation. Acta anat. (Basel) **65**, 594–605 (1966).
27. Boyd, J. D., Hamilton, W. J.: Cleavage, early development and implantation of the egg. In: A. S. Parkes (edit.), Marshall's physiology of reproduction, 3rd edit., vol. 2, p. 1–126. London: Langmans, Green 1952.
28. Calarco, P., Moyer, F. H.: Structural changes in the murine yolk sac during gestation: Cytochemical and electron microscope observations. J. Morph. **119**, 341–356 (1966).
29. Chiquoine, A. D.: The distribution of polysaccharides during gastrulation and embryogenesis in the mouse embryo. Anat. Rec. **129**, 495–516 (1957).
30. Dean, H. W., Rubin, B. L., Dirks, E. C., Lober, B. L., Leipner, G.: Trophoblastic giant cells in placentas of rat and mice and their probable role in steroid hormone production. Endocrinology **70**, 407–419 (1962).
31. Enders, A. C.: A comparative study of the fine structure of the trophoblast in several hemochorial placentas. Amer. J. Anat. **116**, 29–68 (1965).
32. Finn, C. A., McLaren, A.: A study of early stages of implantation in mice. J. Reprod. Fertil. **13**, 259–267 (1967).
33. Friedrich, F.: Die Entwicklung der sogenannten Dottersackdivertikel in der Plazenta der weißen Maus. Z. Anat. Entwickl.-Gesch. **124**, 153–170 (1964).
34. Kirby, D., Bradbury, S.: The hemochorial mouse placenta. Anat. Rec. **152**, 279–282 (1965).
35. Porter, D. G.: Observations on the development of mouse blastocysts transferred to the testis and kidney. Amer. J. Anat. **121**, 73–86 (1967).
36. Reinius, S.: Morphology of the mouse embryo, from the time of implantation to mesoderm formation. Z. Zellforsch. **68**, 711–723 (1965).
37. Reinius, S.: Ultrastructure of blastocyst attachment in the mouse. Z. Zellforsch. **77**, 257–266 (1967).
38. Saccoman, F., Morgan, C. F., Wells, L. J.: Radioautographic studies of DNA-Synthesis in the developing extra-embryonic membranes of the mouse. Anat. Rec. **158**, 197–206 (1967).
39. Schiebler, H.: Mitteilungen über die Riesenzellen der Nagetierplazenta. Verh. Anat. Ges. 54. Verslg. Anat. Anz., Erg.-H. **104**, 318–323 (1958).
40. Smith, L. J.: Metrial gland and other glycogen containing cells in the mouse uterus following mating, and through implantation of the embryo. Amer. J. Anat. **119**, 15–23 (1966).
41. Theiler, K.: Die Anlage des Vorderdarmes bei der Hausmaus, und die Furchenbildungen am Eizylinder. Z. Anat. Entwickl.-Gesch. **128**, 40–46 (1969).

Blood and Circulatory System

42. Albert, S., Wolf, P. L., Potter, R.: Observations on the origin of lymphocyte-like cells in the mouse bone marrow. Nature (Lond.) **212**, 1577–1579 (1966).
43. Andrews, W.: Comparative hematology. New York: Grune & Stratton 1965.
44. Bermann, J.: The ultrastructure of erythroblastic islands and reticular cells in mouse bone marrow. J. Ultrastruct. Res. **17**, 291–313 (1967).
45. Caruso, R., Petrakis, N. L.: Studies of the coagulation and prothrombin time in the mouse embryo. Thrombos. Diathes. haemorrh. (Stuttg.) **16**, 732–737 (1966).
46. Chapelle, A. de la, Fantoni, A., Marks, P. A.: Differentiation of mammalian somatic cells: DNA and hemoglobin synthesis in fetal mouse yolk sac erythroid cells. Proc. nat. Acad. Sci. (Wash.) **63**, 812–819 (1969).
47. Craig, M. L.: A developmental change in hemoglobins, correlated with an embryonic red cell population in the mouse. Develop. Biol. **10**, 191–201 (1964).
48. Orlic, D., Gordon, A. S., Rhodin, J. A. G.: An ultrastructural study of erythropoietin-induced red cell formation in mouse spleen. J. Ultrastruct. Res. **13**, 516–542 (1965).
49. Ravn, E.: Über die Arteria omphalo-mesenterica der Ratten und Mäuse. Anat. Anz. **9**, 420–429 (1894).
50. Salzer, P., Theiler, K.: Über eine besondere Klappe am Ductus venosus Arantii bei der Maus. Z. Anat. Entwickl.-Gesch. **130**, 91–94 (1970).

Lymphnodes, Thymus, Spleen

51. BURNET, F. M.: Mast cells in the mouse thymus. In: G. E. W. WOLSTENHOLME and R. PORTER, The thymus, p. 335–340. Boston: Little & Brown 1966.
52. CRISAN, C.: Die Entwicklung des thyreo-parathyreothymischen Systems der weißen Maus. Z. Anat. Entwickl.-Gesch. **104**, 327–358 (1935).
53. DUNN, T. B.: Normal and pathologic anatomy of the reticular tissue in laboratory mice, with a classification and discussion of neoplasms. J. nat. Cancer Inst. **14**, 1281–1434 (1954).
54. ENGESET, H., TJOTTA, E.: Lymphatic pathways from the tail in rats and mice. Cancer Res. **20**, 613–614 (1960).
55. MANDEL, T.: Differentiation of epithelial cells in the mouse thymus. Z. Zellforsch. **106**, 498–517 (1970).
56. MIYAZAKI, T.: Über die Entwicklung der Lymphknoten bei der Maus. Soc. Path. Jap. Trans. **30**, 29–34 (1940).
57. PRINDULL, G.: Die Milz als nachgeordnetes lymphatisches Organ. Z. Anat. Entwickl.-Gesch. **125**, 255–275 (1966).
58. SANEL, F. T.: Ultrastructure of differentiating cells during thymus histogenesis. Z. Zellforsch. **83**, 8–29 (1967).
59. SANEL, F. T., COPENHAVER, W. M.: Histogenesis of mouse thymus studied with the light and electron microscope. Anat. Rec. **151**, 410 (abstr.) (1965).
60. SMITH, C.: Studies on the thymus of the mammals. XIV. Histology and histochemistry of embryonic and early postnatal thymus of C57BL/6 and AKR strain mice. Amer. J. Anat. **116**, 611–629 (1965).

Palate and Respiratory Tract

61. BROWDER, S.: Factors influencing lung lobation in the mouse. I. Genetic factors. A preliminary report. Anat. Rec. **83**, 31–39 (1942).
62. CALLAS, G., WALKER, B. E.: Palate morphogenesis in mouse embryos after irradiation. Anat. Rec. **145**, 61–68 (1963).
63. CROCKER, T. T., TEETER, A., NIELSEN, B.: Postnatal cellular proliferation in mouse and hamster lung. Cancer Res. **30**, 357–361 (1970).
64. FARBMAN, A. J.: The epithelium-connective tissue interface during closure of the secondary palate in rodent embryos. J. dent. Res. **48**, 617–624 (1969).
65. JACOBS, R. M.: Histochemical study of morphogenesis and teratogenesis of the palate in mouse embryos. Anat. Rec. **149**, 691–698 (1964).
66. KARRER, H. E.: The ultrastructure of mouse lung: The alveolar macrophage. J. biophys. biochem. Cytol. **4**, 693–700 (1958).
67. LARSON, K. S.: Studies on the closure of the secondary palate. Exp. Cell Res. **21**, 498–503 (1960).
68. ROOS, L. M., WALKER, B. E.: Movement of palatine shelves in untreated and teratogen-treated mouse embryos. Amer. J. Anat. **121**, 509–522 (1967).
69. TRASLER, D. G.: Pathogenesis of cleft lip and its relation to embryonic face shape in A/J and C57BL mice. Teratology **1**, 33–50 (1968).
70. WALKER, B. E.: Correlation of embryonic movement with palate closure in mice. Teratology **2**, 191–198 (1968).

Teeth, Oral Cavity

71. BADER, R. S.: Fluctuating asymmetry in the dentition of the house mouse. Growth **29**, 291–300 (1965).
72. CARAMIA, F.: Ultrastructure of mouse submaxillary gland. I. Sexual differences. J. Ultrastruct. Res. **16**, 505–523 (1966).
73. COHN, S. A.: Development of the molar teeth in the albino mouse. Amer. J. Anat. **101**, 295–320 (1957).
74. DISHER, L.: Histogenesis of the mouse submandibular salivary gland. Anat. Rec. **157**, 235 (abstr.) (1967).
75. GAUNT, W. A.: The disposition of the developing cheek teeth in the albino mouse. Acta anat. (Basel) **64**, 572–585 (1966).

76. Hinrichsen, K.: Morphologische Untersuchungen zur Topogenese der mandibularen Nagezähne der Maus. Anat. Anz. **107**, 55–74 (1959).
77. Kelemen, G.: Nonexperimental nasal and paranasal pathology in hereditarily obese mice. Acta oto-laryng. (Stockh.) **57**, 143–151 (1953).
78. Kutuzow, H., Sicher, H.: Comparative anatomy of the mucosa of the tongue and the palate of the laboratory mouse. Anat. Rec. **116**, 409–425 (1953).
79. Weber, J., Kittel, R.: Die Abhängigkeit des Sekretionszyklus der Streifenstücke in der Glandula submandibularis vom ovariellen Zyklus der weißen Maus. Morph. Jb. **108**, 1–17 (1965).

Intestinal Tract

80. Daems, W. T., Wisse, E.: Shape and attachment of the cristae mitochondriales in mouse hepatic cell mitochondria. J. Ultrastruct. Res. **16**, 123–140 (1966).
81. Danforth, C. H., Center, E.: Development and genetics of a sexinfluenced trait in the livers of mice. Proc. nat. Acad. Sci. (Wash.) **39**, 811–817 (1953).
82. Frankenberger, Z.: Sur la morphologie et le développement des voies biliaires chez le genre Mus. Arch. Anat. (Strasbourg) **6**, 201–216 (1926).
83. Green, M.: A defect of the splanchnic mesoderm caused by the mutant gene dominant hemimelia in the mouse. Develop. Biol. **15**, 62–89 (1966).
84. Moog, F.: The functional differentiation of the small intestine. II. The differentiation of alkaline phospho-monoesterase in the duodenum of the mouse. J. exp. Zool. **118**, 187–207 (1951).
85. Overton, J.: Fine structure of the free cell surface in developing mouse intestinal mucosa. J. exp. Zool. **159**, 195–202 (1965).
86. Parakkal, P. F.: An electron microscopic study of esophageal epithelium in the newborn and adult mouse. Amer. J. Anat. **121**, 175–196 (1967).
87. Rutter, W. J., Wessells, N. K., Grobstein, C.: Control of specific synthesis in the developing pancreas. Nat. Cancer Inst. Monogr. **13**, 51–65 (1964).
88. Theiler, K.: Die Anlage des Vorderdarmes bei der Hausmaus und die Furchenbildungen am Eizylinder. Z. Anat. Entwickl.-Gesch. **128**, 40–46 (1969).
89. Wilson, J. W.: Abnormal mitosis in mouse liver. Amer. J. Anat. **86**, 51–74 (1950).
90. Yamada, E.: The fine structure of the gall bladder epithelium of the mouse. J. biophys. biochem. Cytol. **1**, 445–458 (1955).

Urogenital Tract

91. Brown, A. L.: An analysis of the developing metanephros in mouse embryos with abnormal kidneys. Amer. J. Anat. **47**, 117–172 (1931).
92. Crabtree, C.: The structure of Bowman's capsule as an index of age and sex variations in normal mice. Anat. Rec. **79**, 395–413 (1941).
93. Deane, H. W., Wurzelmann, S.: Electron microscopic observations on the postnatal differentiation of the seminal vesicle epithelium of the laboratory mouse. Amer. J. Anat. **117**, 91–134 (1965).
94. Dunn, T. B.: Some observations of the normal and pathologic anatomy of the kidney of the mouse. J. nat. Cancer Inst. **9**, 285–301 (1949).
95. Espinasse, P. G.: The oviduct epithelium of the mouse. J. Anat. (Lond.) **69**, 363–368 (1935).
96. Firlitt, C. F., Davis, J. R.: Morphogenesis of the residual body of the mouse testis. Quart. J. micr. Sci. **106**, 93–98 (1965).
97. Flickinger, C. J.: The postnatal development of the Sertoli cells of the mouse. Z. Zellforsch. **78**, 92–113 (1967).
98. Forsberg, J. G., Olivecrona, H.: Further studies on the differentiation of the epithelium in the mouse vaginal anlage. Z. Zellforsch. **66**, 867–877 (1965).
99. Fuxe, K., Nilsson, O.: The mouse uterine surface epithelium during the estrous cycle. Anat. Rec. **145**, 541–548 (1963).
100. Grobstein, C.: Inductive interaction in the development of the mouse metanephros. J. exp. Zool. **130**, 319–339 (1955).
101. Heinecke, H., Grimm, H.: Untersuchungen zur Öffnungszeit der Vaginalmembran bei verschiedenen Mäusestämmen. Endocrinology **35**, 205–213 (1958).

102. LUDWIG, E.: Beitrag zur Entwicklungsgeschichte der Nachniere. Acta anat. (Basel) **8**, 1–17 (1949).
103. MINTZ, B.: Germ cell origin and history in the mouse: genetic and histochemical evidence. Anat. Rec. **127**, 335–336 (abstr.) (1957).
104. PIERCE, G. B., BEALS, T. F.: The ultrastructure of primordial germinal cells of the fetal testes and of embryonal carcinoma cells of mice. Cancer Res. **24**, 1553–1567 (1964).
105. RAYNAUD, A.: Développement des glandes annexes du tractus génital de la souris. C. R. Soc. Biol. (Basel) **136**, 292–294 (1942).
106. ROSS, M. H.: The fine structure and development of the peritubular contractile cell component in the seminiferous tubules of the mouse. Amer. J. Anat. **121**, 523–558 (1967).
107. WIMSATT, W. A., WALDO, C. M.: The normal occurrence of a peritoneal opening in the bursa ovarii of the mouse. Anat. Rec. **93**, 47–57 (1945).

Spermatogenesis

108. CUTRIGHT, P. R.: Spermatogenesis in the mouse. J. Morph. **54**, 197–220 (1952).
109. EVANS, E. P., BRECKON, G., FORD, C. E.: An air-drying method for meiotic preparations from mammalian testes. Cytogenetics **3**, 289–294 (1964).
110. FOGG, L. C., COWING, R. F.: Cytological changes in the spermatogonial nuclei correlated with increased radioresistance. Exp. Cell Res. **4**, 107–115 (1952).
111. GALTON, M., HOLT, S. F.: Asynchronous replication of the mouse sex chromosomes. Exp. Cell Res. **37**, 111–116 (1965).
112. GARDNER, P. J.: Fine structure of the seminiferous tubule of the Swiss mouse. The spermatid. Anat. Rec. **155**, 235–250 (1966).
113. LAM, D. M. K., FURRER, R., BRUCE, W. R.: The separation, physical characterization and differentiation kinetics of spermatogonial cells of the mouse. Proc. nat. Acad. Sci. (Wash.) **65**, 192–199 (1970).
114. LEBLOND, C. P., CLERMONT, G. C.: Spermiogenesis of rat, mouse, hamster and guinea-pig as revealed by the PAS-technique. Amer. J. Anat. **90**, 167–216 (1952).
115. MONESI, V.: Autoradiographic study of DNA-synthesis and the cell cycle in spermatogonia and spermatocytes of mouse testis using Thymidin H_3. J. Cell Biol. **14**, 1–18 (1962).
116. OAKBERG, E. F.: A description of spermiogenesis in the mouse and its use in analysis of the cycle of the seminiferous epithelium and germ cell renewal. Amer. J. Anat. **99**, 391–414 (1956).
117. OAKBERG, E. F.: Duration of spermatogenesis in the mouse and timing of stages in the cycle of the seminiferous epithelium. Amer. J. Anat. **99**, 507–516 (1956).
118. SAPSFORD, C. S.: Changes in the cells of the sex cords and seminiferous tubules during the development of the testis of the rat and the mouse. Austral. J. Zool. **10**, 178–192 (1962).
119. SAWADA, T.: An electron microscope study of spermatid differentiation in the mouse. Okajimas Folia anat. jap. **30**, 73–80 (1957).
120. SLIZYNSKI, B. M.: A preliminary pachytene chromosome map of the house mouse. J. Genet. **49**, 242–244 (1949).
121. SUGIHARA, R., YASUZUMI, G.: The fine structure of nuclei as revealed by electron microscopy. VI. Relationship between nucleolus formation and sex chromosomes in the type "A" mouse spermatogonia. Z. Zellforsch. **107**, 466–478 (1970).

Oogenesis

122. BEN-OR, S.: Morphological and functional development of the ovary of the mouse. J. Embryol. exp. Morph. **11**, 715–740 (1963).
123. BORUM, K.: Oogenesis in the mouse; a study of the origin of the mature ova. Exp. Cell Res. **45**, 39–47 (1967).
124. KENT, H. A., JR.: Polyovular follicles and multinucleate ova in the ovaries of young mice. Anat. Rec. **137**, 521–524 (1960).
125. MCLAREN, A.: The distribution of eggs and embryos between sides in the mouse. J. Endocr. **27**, 157–181 (1963).
126. ODOR, D. L., BLANDAU, R. J.: Ultrastructural studies on fetal and early postnatal mouse ovaries. Amer. J. Anat. **125**, 177–216 (1969).

Endocrine Glands

127. Carvalheira, A. F., Pearse, A. G. E.: Calcitonin. Symposium on thyreocalcitonin acid C-cells (S. Taylor, ed.), p. 122–126. London: W. Heinemann 1967.
128. Chardard-Ramboult, S.: Développement de la glande thyroide chez la souris pendant la vie intra-utérine. C. R. Soc. Biol. (Paris) **143**, 40–41 (1949).
129. Coupland, R. E.: The post-natal distribution of the abdominal chromaffine tissue in the guinea-pig, mouse and white rat. J. Anat. (Lond.) **94**, 244–256 (1960).
130. Delost, P., Chirvan-Nia, P.: Différences raciales dans l'involution de la zone X surrénalienne chez la souris adulte vièrge. C. R. Soc. Biol. (Paris) **152**, 453–455 (1958).
131. Dunn, T. B.: Ciliated cells of the thyroid of the mouse. J. nat. Cancer Inst. **4**, 555–557 (1944).
132. Ekholm, R.: The ultrastructure of the blood capillaries in the mouse thyroid gland. Z. Zellforsch. **46**, 139–146 (1957).
133. Gorbman, A.: Functional and morphological properties in the thyroid gland, ultimobranchial body, and persisting ductus pharyngo-branchialis IV. of adult mouse. Anat. Rec. **98**, 93–101 (1947).
134. Hitzemann, S. J. W.: Development of enzyme activity in the Leydig cells of mouse testis. Anat. Rec. **143**, 351–361 (1962).
135. Jacobs, B. B.: Variations in thyroid morphology of mice. Proc. Soc. exp. Biol. (N.Y.) **97**, 115–118 (1958).
136. Kerr, T.: The development of the pituitary of the laboratory mouse. Quart. J. micr. Sci. **87**, 3–29 (1946).
137. McPhail, M. K., Read, H. G.: The mouse adrenal. I. Development, degeneration and regeneration of the X-zone. Anat. Rec. **84**, 51–73 (1942).
138. Mirskaia, L., Crew, F. A. E.: Maturity in the female mouse. Proc. Roy. Soc. (Edinb.) **50**, 179–186 (1930).
139. Munger, B. L.: A light and electron microscopic study of cellular differentiation in the pancreatic islets of the mouse. Amer. J. Anat. **103**, 275–312 (1958).
140. Ross, M. H.: Fine structure of the juxta-medullary region of the mouse adrenal cortex with special reference to the X-zone. Anat. Rec. **157**, 313 (abstract) (1967).
141. Sano, M., Sasaki, F.: Embryonic development of the mouse anterior pituitary studied by light and electron microscopy. Z. Anat. Entwickl.-Gesch. **129**, 195–222 (1969).
142. Sato, T., Jshikawa, K., Aoi, T., Kitoh, J., Sugiyama, S.: Electron microscopic observations on the development of the parafollicular cells from the ultimobranchial cyst in the thyroid gland of the mouse. Okajimas Folia anat. jap. **42**, 91–105 (1966).
143. Smith, C., Clifford, C. P.: Histochemical study of aberrant thyroid glands associated with the thymus of the mouse. Anat. Rec. **143**, 229–238 (1962).
144. Smithberg, M., Bbunner, M. N.: The induction and maintenance of pregnancy in prepuberal mice. J. exp. Zool. **133**, 441–457 (1956).
145. Stoeckel, M. E., Porte, A.: Observations ultrastructurales sur la parathyroide de souris. I. Etude chez la souris normale. Z. Zellforsch. **73**, 488–502 (1966).
146. Treilhou-Lahille, F., Zylberberg-Jeanmaire, R.: Histologie topographique des dérivés pharyngiens de l'embryon de souris C57/BL entre le 12e et le 15e jour de gestation. C. R. Ass. Anat. 53e Congr., No 144, 1828–1834 (1969).
147. Van Heyningen, H. E.: The initiation of thyroid function in the mouse. Endocrinology **69**, 720–727 (1961).
148. Von Bartheld, F., Moll, J.: The vascular system of the mouse epiphysis with remarks on the comparative anatomy of the venous trunks in the epiphyseal area. Acta anat. (Basel) **22**, 227–235 (1954).
149. Waring, H.: The development of the adrenal gland of the mouse. Quart. J. micr. Sci. **78**, 329–366 (1935).

Brain, Sensory Organs, Skin

150. Wessells, N. K., Roessner, K. D.: Non-proliferation in dermal condensations of mouse vibrissae and pelage hairs. Develop. Biol. **12**, 419–433 (1965).
151. Billingham, R. E., Silvers, W. K.: The melanocytes of mammals. Quart. Rev. Biol. **35**, 1–40 (1960).

152. CHASE, H. B.: The behaviour of pigment cells and epithelial cells in the hair follicle. In: The biology of hair growth (chap. I). New York: Academic Press 1958.
153. CHASE, H. B., MANN, S. J.: Phenogenetic aspects of some hair and pigment mutants. J. cell. comp. Physiol. **56**, 103–111 (1960).
154. CHIQUOINE, A. D.: Distribution of alkaline phosphomono-esterase in the central nervous system of the mouse embryo. J. comp. Neurol. **100**, 415–440 (1954).
155. COHEN, A. J.: Electron microscopic observations of the developing mouse eye. I. Basement membranes during early development and lens formation. Develop. Biol. **3**, 297–316 (1961).
156. DE LONG, B. R., SIDMAN, R. L.: Effects of eye removal at birth on histogenesis of the mouse superior colliculus: An autoradiographic analysis with tritiated Thymidine. J. comp. Neurol. **118**, 205–224 (1962).
157. FAWCETT, D. W.: A comparison of the histological organization and cytochemical reactions of brown and white adipose tissue. J. Morph. **90**, 363–406 (1952).
158. GRAUMANN, W.: Zelldegeneration im Telencephalon medium und Paraphysenentwicklung bei der weißen Maus. Z. Anat. Entwickl.-Gesch. **115**, 19–31 (1950).
159. HANNA, C.: Changes in DNA, RNA and protein synthesis in the developing lens. Invest. Ophthal. **4**, 480–495 (1965).
160. HERRLINGER, H.: Licht- und elektronenmikroskopische Untersuchungen am Subcommissuralorgan der Maus. Ergebn. Anat. Entwickl.-Gesch. **42**, H. 5 (1970).
161. KALLEN, B.: Notes on the development of the neural crest in the head of Mus musculus. J. Embryol. exp. Morph. **1**, 393–398 (1953).
162. KALLEN, B., LINDSKOG, B.: Formation and disappearance of neuromery in Mus musculus. Acta anat. (Basel) **18**, 273–282 (1953).
163. KANEKO, K.: Über die Entwicklung der Thalamuskerne der Maus. Folia anat. jap. **19**, 557–596 (1940).
164. KOVAC, W., DENK, H.: Der Hirnstamm der Maus. Wien-New York: Springer 1968.
165. KNUDSEN, P. A.: Mode of growth of the chorioid plexus in mouse embryos. Acta anat. (Basel) **57**, 172–182 (1964).
166. MATTANZA, G.: Über die Bedeutung des embryonalen Zelluntergangs im Vorderhirn. Acta anat. (Basel) (1972) (im Druck).
167. MARSDEN, H. M., SILVERS, W. K.: The effects of genotype and cell environment on melanoblast differentiation in the house mouse. Genetics **41**, 429–450 (1956).
168. MELARAGNO, H. P., MONTAGNA, W.: The tactile hair follicles in the mouse. Anat. Rec. **115**, 129–150 (1953).
169. MELLER, K., BREIPOHL, W., GLEES, P.: Ontogeny of the mouse motor cortex. The polymorph layer or layer VI. A Golgi and electron-microscopic study. Z. Zellforsch. **99**, 443–458 (1969).
170. MENEFEE, M. G.: The differentiation of keratin-containing cells in the epidermis of embryo mice. Anat. Rec. **122**, 181–191 (1953).
171. MOYER, F. H.: Electron microscope studies on the origin, development and genetic control of melanin granules in the mouse eye. In: G. K. SMELSER, The structure of the eye. New York: Academic Press 1961.
172. NIIMI, K., HARADA, I., KUSAKA, Y., KISHI, S.: The ontogenetic development of the diencephalon in the mouse. Tokushima J. exp. Med. **8**, 203–238 (1961).
173. PIERCE, E. T.: Histogenesis of the nuclei grisei pontis, corporis pontobulbaris and reticularis tegmenti pontis (Bechterew) in the mouse. J. comp. Neurol. **126**, 219–240 (1966).
174. RAWLES, M. E.: Origin of pigment cells from the neural crest in the mouse. Physiol. Zool. **20**, 248–266 (1947).
175. RICHARDSON, F. L.: The acinar pattern in the mammary glands of virgin mice at different ages. J. nat. Cancer Inst. **38**, 305–315 (1967).
176. SEIFERT, K., ULE, G.: Die Ultrastruktur der Riechschleimhaut der neugeborenen und jugendlichen weißen Maus. Z. Zellforsch. **76**, 147–169 (1967).
177. SIDMAN, R. L.: Histogenesis of mouse retina. Studies with Thymidine-H3. In: The structure of the eye (G. SMELSER). New York: Academic Press 1961.
178. TURNER, C. W., GOMEZ, E. T.: The normal development of the mammary gland of the male and female albino mouse. Univ. Miss. Res. Bull. 182 (1932).
179. UZMANN, L.: The histogenesis of the mouse cerebellum as studied by its tritiated thymidine uptake. J. comp. Neurol. **114**, 137–159 (1960).

180. WEIBEL, E. R.: Zur Kenntnis der Differenzierungsvorgänge im Epithel des Ductus cochlearis. Acta anat. (Basel) **29**, 53–90 (1957).
181. WRENN, J. T., WESSELLS, N. K.: An ultrastructural study of lens invagination in the mouse. J. exp. Zool. **171**, 359–368 (1969).

Skeletal System, Musculature

182. BATEMAN, N.: Bone growth: A study of the grey-lethal and microphthalmic mutants in the mouse. J. Anat. (Lond.) **88**, 212–260 (1954).
183. CHEN, J. M.: Studies on the morphogenesis of the mouse sternum. I. Normal embryonic development. J. Anat. (Lond.) **86**, 373–386 (1952).
184. FORSTHOEFEL, P. F.: Observations on the sequence of blastemal condensations in the limbs of the mouse embryo. Anat. Rec. **147**, 129–138 (1963).
185. FROMMER, J.: Prenatal development of the mandibular joint in mice. Anat. Rec. **150**, 449–461 (1964).
186. GRUNEBERG, H.: The pathology of development. A study of inherited skeletal disorders in animals, 390 p. Oxford: Blackwell 1963.
187. HOSHINO, K.: Comparative study on the skeletal development in the fetus of rat and mouse. Congenit. Anomalies (Japan) **7**, 32–38 (1967).
188. JOHNSON, M. L.: The time and order of appearance of ossification centers in the albino mouse. Amer. J. Anat. **52**, 241–271 (1933).
189. THEILER, K.: Zelluntergang in den hintersten Rumpfsomiten bei der Maus. Z. Anat. Entwickl.-Gesch. **120**, 274–278 (1958).
190. THEILER, K.: Experimentelle Segmentierungsstörungen. Verh. Anat. Ges. (Marburg 1967). Erg.-H. Anat. Anz. **121**, 557–560 (1968).
191. WIRSEN, C., LARSSON, K. S.: Histochemical differenciation of skeletal muscle in foetal and newborn mice. J. Embryol. exp. Morph. **12**, 759–767 (1964).
192. WIRTSCHAFTER, Z. T.: Genesis of the mouse skeleton. Springfield: Ch. C. Thomas 1966.

Growth

193. ATLAS, M., BOND, V. P.: The cell generation cycle of the eleven-day mouse embryo. J. Cell Biol. **26**, 19–24 (1965).
194. GOEDBLOED, J. F.: La croissance de l'embryon du rat et de la souris, et quelques-uns de leurs organes, et la croissance post-natale de la souris. C. R. Ass. Anat. 53e Congr., No 142, 933–941 (1969).
195. GRUNEBERG, H.: A ventral ectodermal ridge of the tail in mouse embryos. Nature (Lond.) **177**, 787–788 (1956).
196. LAIRD, A. K., HOWARD, A.: Growth curves in inbred mice. Nature (Lond.) **213**, 786–788 (1967).
197. MCDOWELL, E. C., ALLEN, E., MCDOWELL, C. G.: The prenatal growth of the mouse. J. gen. Physiol. **11**, 57–70 (1927).
198. MCDOWELL, E. C., GATES, W. H., MCDOWELL, C. G.: The influence of the quantity of nutrition upon the growth of the suckling mouse. J. gen. Physiol. **13**, 529–545 (1930).
199. MCLAREN, A.: Genetics and environmental effects on foetal and placental growth in mice. J. Reprod. Fertil. **9**, 79–98 (1965).
200. OGANA, T.: Comparative study on development in the stage of organogenesis in the mouse and rat. Congenit. Anomalies (Japan) **7**, 27–31 (1967).
201. READING, A. J.: Influence of room temperature on the growth of house mice. J. Mammal. **47**, 694–697 (1966).
202. SCHUMANN, H.: Vergleichende Untersuchungen über das Wachstum von Ratten- und Mäuse-embryonen. Wilhelm Roux' Archiv **163**, 325–333 (1969).

Addenda

Opera Generalia

OLIVIER, G., PINEAU, H.: Horizons de Streeter et âge embryonnaire. C. R. Ass. Anat. 47e Réun. Naples 1961, **113**, 573–576 (1962).

Pre-implantation Period

Bergström, S.: Shedding of the zona pellucida in normal pregnancy and in various hormonal states in the mouse. A scanning electron microscope study. Z. Anat. Entwickl.-Gesch. **136**, 143–167 (1972).

Cholewa, J. A., Whitten, W. K.: Development of two-cell mouse embryos in the absence of a fixed nitrogen source. J. Reprod. Fertil. **22**, 553–555 (1970).

Stern, S., Biggers, J. D., Anderson, E.: Mitochondria and early development of the mouse. J. exp. Zool. **176**, 179–192 (1971).

Implantation and Fetal Membranes

Solter, D., Damjanov, I., Skreb, N.: Ultrastructure of mouse egg-cylinder. Z. Anat. Entwickl.-Gesch. **132**, 291–298 (1970).

Blood and Cardio-vascular System

Bugge, J.: The contribution of the stapedial artery to the cephalic arterial supply in muroid rodents. Acta anat. (Basel) **76**, 313–336 (1970).

Jones, R. O.: Ultrastructural analysis of hepatic haematopoiesis in the foetal mouse. J. Anat. (Lond.) **107**, 301–314 (1970).

Mierop, L. H. S. van, Gessner, J. H.: The morphologic development of the sinoatrial node in the mouse. Amer. J. Cardiol. **25**, 204–212 (1970).

Lymphnodes, Thymus

Abe, K., Ito, T.: Fine structure of small lymphocytes in the thymus of the mouse. Qualitative and quantitative analysis by electron microscope. Z. Zellforsch. **110**, 321–335 (1970).

Respiratory Tract

Hatasa, K., Nakamura, T.: Electron microscopic observations of lung alveolar epithelial cells of normal young mice with special reference to formation and secretion of osmiophilic lamellar bodies. Z. Zellforsch. **68**, 266–277 (1965).

Woodside, G. L., Dalton, A. J.: The ultrastructure of lung tissue from newborn and embryo mice. J. Ultrastruct. Res. **2**, 28–54 (1958).

Oral Cavity

Berkman, M. D., Kronman, J. H.: A histochemical study of the effects of castration and testosterone administration of the major salivary glands of Swiss mice. Acta anat. (Basel) **76**, 200–219 (1970).

Yohro, T.: Development of secretory units of mouse submandibular gland. Z. Zellforsch. **110**, 173–184 (1970).

Urogenital Tract

Forsberg, J.-G., Åbro, A.: Ultrastructural differences between the sinus and the Müllerian epithelium of the mouse vaginal anlage. Z. Anat. Entwickl.-Gesch. **135**, 67–75 (1971).

Mullen, R. J., Whitten, W. K.: Relationship of genotype and degree of chimerism in coat color to sex ratios and gametogenesis in chimeric mice. J. exp. Zool., in press (1971).

Odor, D. L., Blandau, R. J.: Ultrastructural studies on fetal and early postnatal mouse ovaries. II. Cytodifferentiation. Amer. J. Anat. **125**, 177–216 (1969).

Intestinal Tract

Yamada, K.: The glands of the common bile duct in the rat and mouse. An electron microscope study. Acta anat. (Basel) **77**, 438–453 (1970).

Endocrine Glands

BROECKER, E. B.: Die Lipoidtropfen in der Nebennierenrinde der weiblichen Maus und Beziehungen ihrer Größenverteilungen zum Zyklus. Z. Zellforsch. **113**, 188–202 (1971).

EURENIUS, L., JARSKAR, R.: Electron microscope studies on the development of the external zone of the mouse median eminence. Z. Zellforsch. **122**, 488–502 (1971).

FERNHOLM, M.: On the development of the sympathetic chain and the adrenal medulla in the mouse. Z. Anat. Entwickl.-Gesch. **133**, 305–317 (1971).

GARWEG, G., KINSKY, J., BRINKMANN, H.: Markierung der juxtamedullären X-Zone in der Nebenniere der Maus mit L-Cystein-S^{35}. Z. Anat. Entwickl.-Gesch. **134**, 186–199 (1971).

STOECKEL, M. E., DELLMANN, H. D., PORTE, A., GERTNER, C.: The rostral zone of the intermediate lobe of the mouse hypophysis, a zone of particular concentration of corticotrophic cells. Z. Zellforsch. **122**, 310–322 (1971).

Brain, Sensory Organs, Skin

DRY, F. W.: The coat of the mouse. J. Genet. **16**, 287–340 (1926).

HINDS, J. W., RUFFETT, T. L.: Cellproliferation in the neural tube: An electron microscopic and Golgi analysis in the mouse cerebral vesicle. Z. Zellforsch. **115**, 226–264 (1971).

SIDMAN, R. L., ANGEVINE, J. B., PIERCE, E. T.: Atlas of the mouse brain and spinal cord. Berlin-Heidelberg-New York: Springer 1971.

Growth

YAMAMURA, H.: Individuelle Unterschiede des Entwicklungsstandes bei Embryonen der Maus (C57BL) in der frühen Phase der Organogenese. Wilhelm Roux' Archiv **162**, 218–242 (1969).

Subject Index